AEROSOLS - SCIENCE AND CASE STUDIES

Edited by **Konstantin Volkov**

Aerosols - Science and Case Studies
http://dx.doi.org/10.5772/62964
Edited by Konstantin Volkov

Contributors
Yu Feng, Zelin Xu, Ahmadreza Haghnegahdar, Mingzhou Yu, Zahary Yordanov Peshev, Gajanan Aher, Daouda Diouf, Awa Niang, Sylvie Thiria, Julien Brajard, Salam Sawadogo, Michel Crépon, Zhihua Mao, Kun Zhou, Tat Leung Chan, Kyriacos Themistocleous

Published by InTech
Janeza Trdine 9, 51000 Rijeka, Croatia

Publishing Process Manager Maja Bozicevic
Technical Editor SPi Global
Cover InTech Design team

First published December, 2016
Printed in Croatia

Additional hard copies can be obtained from orders@intechopen.com

Aerosols - Science and Case Studies, Edited by Konstantin Volkov
p. cm.
Print ISBN 978-953-51-2843-4
Online ISBN 978-953-51-2844-1

For EU product safety concerns:
IN TECH d.o.o., Prolaz Marije Krucifikse Kozulić 3, 51000 Rijeka, Croatia,
info@intechopen.com or visit our website at intechopen.com.

Meet the editor

Dr. Volkov is a senior lecturer in Thermofluids at the Kingston University (London, UK). He holds a PhD in fluid mechanics. After completion of his PhD, Dr. Volkov worked at the Baltic State Technical University (Russia), University of Central Lancashire (UK) and University of Surrey (UK). His areas of expertise cover multidisciplinary areas from design and optimization of energy systems to fundamental problems focused on modelling and simulation. He is the Chartered Engineer and a member of Institute of Physics, Institution of Mechanical Engineers and Combustion Institute. He is the author of more than 120 scientific papers and a member of editorial board and scientific committee of a number of journals and conferences.

Contents

Preface

An aerosol is a type of two-phase mixture containing fine (nano- or micro-scale) solid particles or liquid droplets in the air. Aerosols can be natural or artificial. Typical example of natural aerosols is fog, and examples of artificial aerosols include dust, particulate air pollutants and smoke. Aerosols are often produced through a complex interaction of sunlight; volatile organic compounds from trees, plants and cars or industrial emissions; and other airborne chemicals.

Aerosol particles and droplets have a significant impact on technological processes and human activities. The role of particles and droplets in environmental and industrial applications is twofold. On the one hand, particles and droplets are successfully used in engineering solutions. On the other hand, particles and droplets may pose potential hazard. In many cases, aerosols are at the core of human health, environmental and technological problems (climate change and air quality). However, aerosols can be successfully used in industry and technology (new materials, fire suppression and fuel delivery). Processes that control transport and dynamics of aerosol particles and droplets remain unresolved and introduce significant uncertainties into modelling and simulation.

The current scientific status of aerosol modelling and simulation and measurements and some advances in computational techniques, particle measurement technologies and practical applications of aerosols are reviewed and considered in this book. This book also includes a number of case studies focused on analysis of optical thickness and air quality in various regions.

The book contains eight chapters written by specialists from various countries who are working in various fields related to predictions of aerosol properties and their impact on human health or use of aerosols in industry.

Operator Splitting Monte Carlo Method for Aerosol Dynamics

Different methods are used in practice to solve the population balance equation describing aerosol dynamics, in particular direct discretization, method of moments and stochastic method. Being the most flexible among other methods, the stochastic method is time-consuming and expensive from computational point of view. An operator splitting Monte Carlo (OSMC) method was developed to improve the numerical efficiency while preserving the flexibility of the stochastic method. Nucleation and surface growth are handled with deterministic means, while coagulation is simulated with a stochastic method (Marcus-Lushnikov stochastic process). The operator splitting errors of various schemes are analysed, and some numerical examples are provided.

Methods of Moments for Resolving Aerosol Dynamics

The solution of the population balance equation describing the formation of nano- and micro-scale aerosol particles and their subsequent growth in quiescent or evolving flows has received significant attention from chemical engineers and atmospheric environment community for the past years. Development of the method of moments to solve the population balance equation is presented. Three main methods of moments, including the Taylor-series expansion method, log-normal method and quadrature method, are discussed and analysed.

Computational Fluid-Particle Dynamics Modelling for Unconventional Inhaled Aerosols in Human Respiratory Systems

Nano- and micro-scale aerosol particles which are mostly non-spherical and hygroscopic are used in medicine and pharmaceutical industry. Fundamentals and future trends of computational fluid-particle dynamics models for lung aerosol dynamics are reviewed and discussed with emphasis on physics of unconventional inhaled aerosols. Reconstruction of the human respiratory system, numerical model formulation and mesh generation are described. The case studies for fibre and droplet transport and deposition in the lung are also provided. The study contributes to development of multi-scale numerical model allowing to simulate complex airflow-vapour-particle-structure dynamics of the entire human respiratory system in detail.

Lidar Mapping of Near-Surface Aerosol Fields

Aerosol pollutions in the low troposphere have a significant impact on human health. Near-surface atmospheric measurements over urban or industrial areas and monitoring of air quality become important problems. The experimental results on lidar mapping of aerosol fields over the Sofia city (Bulgaria) obtained during seven-month experimental campaign in 2015 are presented. The measurements are conducted by scanning observation zones in horizontal and vertical directions. Two-dimensional colour-coded sector maps of the near-surface aerosol density are obtained using the aerosol backscattering profiles retrieved at different azimuth or elevation angles. The analysis of the lidar maps shows a good correlation between the aerosol density distribution and the locations of important sources of aerosol pollutions (city streets with intensive traffic, industrial facilities and densely populated residential districts). The results reported demonstrate that aerosol lidar mapping could be regarded as an effective approach for accurate and reliable determination of the density, spatial distribution and temporal dynamics of close-to-ground aerosols, covering broad urban areas.

Aerosols Monitored by Satellite Remote Sensing

Aerosol particles suspended in atmosphere affect air quality and climate change. Global distribution of aerosol particles is monitored by satellite remote sensing. The Cloud-Aerosol Lidar and Infrared Pathfinder Satellite Observation (CALIPSO) satellite provides global, vertically resolved measurements of aerosol distribution and aerosol extinction coefficients. This allows to perform height-resolved discrimination of aerosol into several types providing a new insight into the role that atmospheric aerosols play in regulating the Earth's weather, climate and air quality.

The Image-Based Integrated Method for Determining and Mapping Aerosol Optical Thickness

An image-based integrated method for determining and mapping aerosol optical thickness is developed. The methodology proposed uses the radiative transfer equation and non-variant targets with the revised darkest pixel method to improve the accuracy of predictions of optical thickness in urban areas. The method developed is applied to Landsat TM/ETM+ satellite images of Limassol (Cyprus) over a period of time. An accuracy assessment of the method shows a strong correlation between the values from the sun photometers and the values derived from the image-based integrated method.

Spatial Distribution of Aerosol Optical Thickness Retrieved from SeaWiFS Images by a Neural Network Inversion over the West African Coast

The desert aerosol plumes characterized by high optical thicknesses often crossing West African coast affect accuracy of optical thickness predictions. Aerosol optical thickness predictions from December 1997 to November 2009 by using neural networks inversion make possible a better understanding of climate change in a specific region.

Aerosol-Cloud Interaction: A Case Study

Aerosol and cloud interactions are critical in understanding the climate change since clouds play an important role in controlling incoming and outgoing radiation. Extensive studies have been performed for the past years highlighting various mechanisms of cloud properties through the interaction of atmospheric aerosol particles with cloud parameters. The study deals with the spatial and temporal variations of the MODIS retrieved aerosol and cloud products at the nine selected stations over Western Himalaya and Deccan Plateau regions. Western Himalaya is desert dust–dominated region, while Deccan Plateau is dominated by fossil fuel and biomass burning. The data obtained allow to investigate aerosol-cloud interaction and to quantify the aerosol indirect effect over these regions.

This book covers many aspects of aerosol science representing the latest research of various groups of internationally recognized experts. This book is intended for engineers and technical workers whose work is related to predictions of aerosol properties. It will be of interest to academics working in environmental engineering and to industrial practitioners in companies concerned with design and optimisation of systems with solid or liquid particles. The works presented in the book are easily extendible to be relevant in other areas in which aerosols play an important role.

The open exchange of scientific data, results and ideas will hopefully lead to improved predictions of aerosol impact on human activities and industrial processes. This book presents necessary data and helpful suggestions to advance understanding the formation, atmospheric fate and transport, health effects and climate change interactions of aerosols.

Dr. Konstantin Volkov
MEng, MSc, PhD, DSc, CEng, MIMechE, MInstP
Department of Mechanical and Automotive Engineering
School of Mechanical and Aerospace Engineering
Faculty of Science, Engineering and Computing
Kingston University
London, United Kingdom

Chapter 1

Operator Splitting Monte Carlo Method for Aerosol Dynamics

Kun Zhou and Tat Leung Chan

Additional information is available at the end of the chapter

http://dx.doi.org/10.5772/65140

Abstract

Aerosol dynamics are described by the population balance equation (PBE). In principle, three typical methods (i.e., direct discretization, method of moments, and stochastic method) have been widely used to solve the PBE. Stochastic method is the most flexible among the three methods. However, stochastic method is computationally expensive. Recently, an operator splitting Monte Carlo (OSMC) method has been developed so as to improve the numerical efficiency while preserving the flexibility of the stochastic method. Within the OSMC, nucleation and surface growth are handled with deterministic means, while coagulation is simulated with a stochastic method (the Marcus-Lushnikov stochastic process). The stochastic and deterministic treatments of various aerosol dynamic processes are synthesized under the framework of operator splitting. Here, the operator splitting errors of various schemes are analyzed rigorously, combined with concrete numerical examples. The analyses not only provide sound theoretical bases for selecting the most efficient operator splitting scheme for the usage of the OSMC, but also shed lights on how to adopt operator splitting in other PBE solving methods, i.e., direct discretization, method of moments, etc.

Keywords: Monte Carlo simulation, aerosol dynamics, operator splitting

1. Introduction

Aerosols are dispersed systems of solid or liquid particles suspended in air. Aerosols are created during various natural processes and industrial operations [1], such as clouds, smoke particles from forest fires, sand and dust storms, volcanic dust, spores and seeds from plant life, fluidized catalysts, spray drying of fluids, atomization of liquid fuels, fly ash from stacks, soot in combustion flames, etc. Aerosol evolution is very complex, involving various physical and chemical processes. The most popular description of aerosol evolution is using the population balance equation (PBE) [2], which is also called the general dynamic equation in aerosol field [3]. The PBE is a convection-diffusion type equation with source terms taking into account the particle formation and depletion from various aerosol dynamics, mainly nucleation, surface growth (condensation), and coagulation.

Analytical solutions to the PBE are available for only a few specific cases [4–7]. Mostly, numerical means are resorted to solve the PBE. There are generally three typical methods for solving the PBE, i.e., direct discretization, method of moments, and stochastic methods. Direct discretization may be in the particle size space, such as discretization by sections [8] or finite difference/finite element [9] and may also be in the functional space, such as those in the collocation method [10], Galerkin method [11], and weighted residual method [12]. Among the various discretizations, the sectional method pioneered by Gelbard et al. [8] is the most popular. There are quite a few sectional methods developed aiming to alleviate the nonpositiveness and nonconservation problems found in earlier works [13], and to obtain better efficiency and flexibility, such as the fixed/moving pivot method [14, 15] and cell average method [16, 17].

Other than solving the discretized PBE, a widely used alternative approach is the method of moments, which solves a group of moments equations derived from the PBE. These moments (generally several low order moments) provide important information on the particle size distribution (PSD) function, i.e., number density, volume fraction, polydispersity, etc. However, owing to the nonlinearity of the PBE, the governing equations for lower order moments generally contain higher order moments, which are not closed after cutting off at a certain order of moment. Various closure methods have been developed, such as unimodal log-normal approximation [18], quadrature/direct quadrature [19, 20], interpolative closure [21], Taylor expansion [22], etc. Although the method of moments is computationally highly efficient, the complexity of the closure makes it difficult to accommodate complex physical models for aerosol dynamics with high flexibility. Another issue closely tied to method of moments is the realizability problem, when the moments sequence fails to find its corresponding distribution function [23, 24]. Among all the drawbacks, the inability to provide the PSD is the most severe, when the PSD is highly needed.

Stochastic simulation (or Monte Carlo simulation) does not solve the PBE, but to mimic the evolution of aerosol particles through a stochastic particle system. Based on the Marcus-Lushnikov process modeling of coagulation [29], various stochastic methods have been developed to simulate the general aerosol evolution. It is very difficult to categorize precisely so many diversified methods. According to the evolution of time, there is time-driven or event-driven Monte Carlo [30, 31]. According to the different ways to weight the numerical particles, there are the mass flow algorithms [32, 33], differentially weighted algorithm [34], and the more general weighted flow algorithms [35]. Also, there are various types of methods to improve the simulation efficiency, such as the majorant kernel method [36] and the τ leaping method [37].

Among the three typical methods, stochastic simulation is the most flexible for accommodating complex physical models for various aerosol dynamics, and is easy to implement without special numerical issues, which are often encountered in the other two methods, e.g., nonpositiveness and nonconservation problems in direct discretization, realizability problem in method of moments. However, stochastic simulation generally has convergence rate proportional to $\sqrt{N}$, where N is the number of total numerical particles used. The simulation cost (CPU time and memory) is usually proportional to N^2 [25]. Hence, stochastic simulation is not numerically efficient when pursuing a high precision solution. Recently, Zhou et al. [26] developed an operator splitting Monte Carlo (OSMC) method, which combines stochastic simulation for coagulation process and deterministic integration for nucleation and surface growth (condensation). Stochastic simulation for coagulation is much more efficient than directly solving the Smoluchowski integro-differential equation (the coagulation model equation) with traditional discretization [27]. On the other hand, deterministic integration for nucleation and surface growth processes is far more efficient than stochastic simulations. The stochastic simulation and the deterministic integration in OSMC are synthesized under the framework of operator splitting. The accuracy and efficiency of the OSMC have been demonstrated through various testing cases [26]. Especially, the OSMC has been implemented with Message Passing Interface (MPI) to achieve very promising parallel simulation efficiency [26]. Here, the impact of various operator splitting schemes on the accuracy and efficiency of the OSMC is addressed.

2. Methodology

2.1. Governing equations

Aerosol particles are generally described statistically in a mesoscopic scale through defining the PSD, $n(\vec{\xi},\vec{x},t)$, and $n(\vec{\xi},\vec{x},t)d\xi$ denotes the number density of aerosol particles at location $\vec{x}$ at time t with aerosol properties between $\vec{\xi}$ and $\vec{\xi}+d\vec{\xi}$. Here, $\vec{\xi}$ is usually called an internal variable, which can be volume, surface, constituent, etc. For simplicity, it is assumed $\vec{\xi}$ to be the particle size (volume or diameter specified by the context). It is worth pointing out that the OSMC method can also deal with the cases with a vector type $\vec{\xi}$, i.e., combined properties to describe particles. In fact, the current OSMC code treats the diameter, surface, and volume of an aerosol particle independently, which can easily model particles of various morphologies, such as particles with fractal dimensions. The evolution of the PSD function satisfies the PBE (also called the general dynamic equation) [3]

$$\frac{\partial n}{\partial t}+\nabla\cdot n\vec{u}=\nabla\cdot D\nabla n+I_{\mathrm{nuc}}+G_{\mathrm{cond}}+C_{\mathrm{coag}}, \tag{1}$$

where $\vec{u}$ is the gas flow velocity, and D is the diffusion coefficient. The source terms of I_{nuc}, G_{cond}, and C_{coag} on the right hand side denote nucleation, surface growth (condensation), and coagulation, respectively. The determination of the convection velocity $\vec{u}$ requires solving the fluid dynamics control equations, and the coupling of the fluid dynamics with the aerosol dynamics is another important research field. Here, the gas velocity $\vec{u}$ is assumed known, using some techniques (often computational fluid dynamics). It is worth pointing out that the coupling between the fluid dynamics and the aerosol dynamics is generally handled by the operator splitting method [11, 28–30], i.e., to solve the fluid dynamics and the aerosol dynamics one by one with the feedback from the other dynamics temporarily freezing, and to iterate between these two dynamics to obtain a converged solution.

Another simplification in this work is to neglect the diffusion term in the PBE. For aerosol particles, the mass diffusion due to the number concentration gradient ($\nabla \cdot D \nabla n$) is usually very small compared to the air viscous diffusion, which can be described by the Schmidt number (i.e., the ratio of the viscous diffusion to the mass diffusion). For example, under standard atmospheric conditions, a spherical particle with a diameter of 10 nm has a Schmidt number equal to 290, and a 100 nm particle has Schmidt number 2.2×10^4 [3]. So it is legitimate to neglect the aerosol mass diffusion under very general conditions.

With the assumptions that the gas velocity $\vec{u}$ is known and the mass diffusion term is negligible, the PBE can be simplified as

$$\frac{\partial n}{\partial t^*} = I_{\text{nuc}} + G_{\text{cond}} + C_{\text{coag}}. \tag{2}$$

The convection term on the left hand side of the original PBE (1) has been implicitly solved with the introduction of the Lagrangian time t^*

$$t^*(\vec{x}) = \int_0^{\vec{x}} \frac{d\vec{x}'}{\vec{u}(\vec{x}')}, \tag{3}$$

The transformation converts the Eulerian point of view in Eq. (1) to the Lagrangian point of view in Eq. (2), with the help of the assumed known velocity field $\vec{u}$. There are many practically interested conditions, where the flow field can be considered as steady and one-dimensional, e.g., burner stabilized flames and counter flow diffusion flames. Then it is very easy to integrate Eq. (3) to build up a one-to-one mapping between the time and the spatial coordinate. Hence, the convection can be readily assimilated in the model Eq. (2). For general cases of aerosol evolution in three dimensional flows, it is possible to use the Lagrangian particle scheme to simulate the aerosol evolution along Lagrangian trajectories [31, 32]. So the model Eq. (2) and the subsequent OSMC method may have very broad applications. Later on, t^* is simply denoted as t, since the Lagrangian point of view is not explicitly referred any longer.

The aerosol dynamic processes on the right hand side describe the interactions among molecules (or cluster of molecules) and particles. Nucleation is the process that dozens or hundreds of molecules form a stable critical nucleus. Surface growth includes physical

http://dx.doi.org/10.5772/65140

condensation and surface chemical reactions, which involves interactions between gas phase molecules and an aerosol particle. Surface chemical reactions will not be discussed here due to their vastness. Coagulation is the process that two particles coalesce to form a new one. More details about the aerosol dynamic processes are presented in Section 2.2.

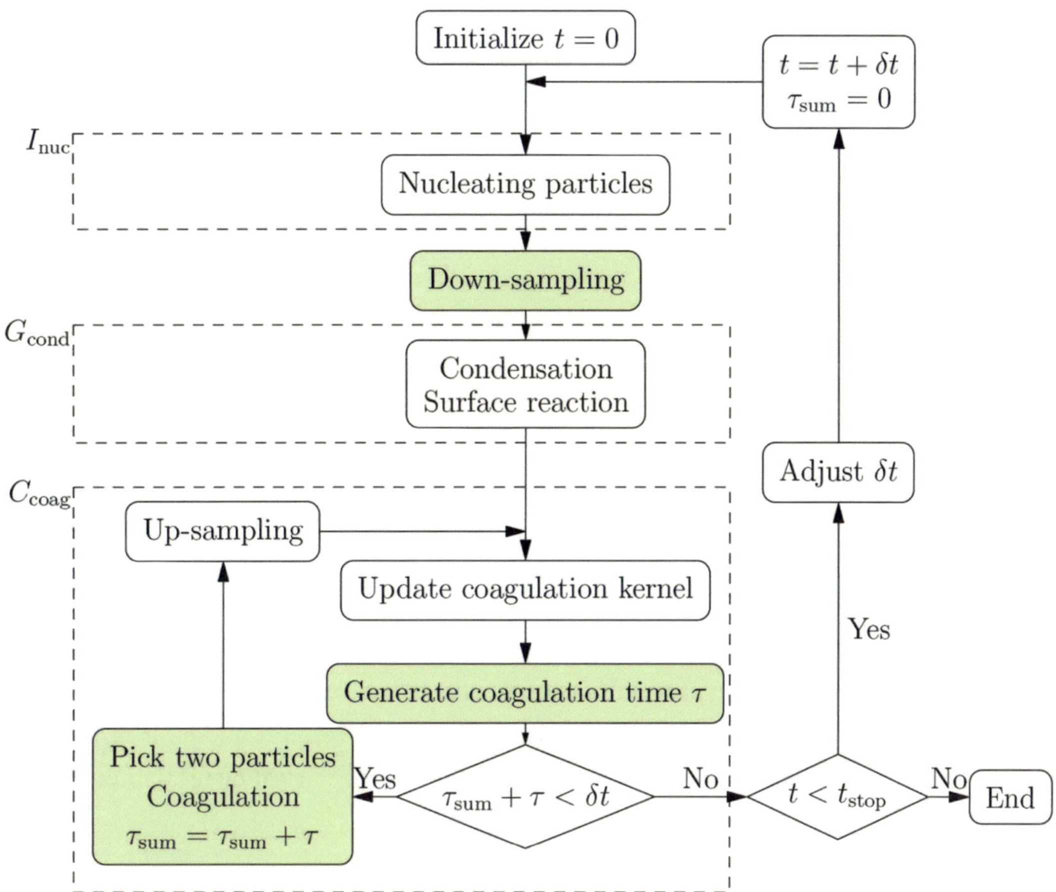

Figure 1. Flowchart of the OSMC. Filled blocks denote steps where stochasticity arises. Adapted from Reference [26].

2.2. Operator splitting Monte Carlo method

Operator splitting is an efficient and powerful method to solve such evolution equations [33] as Eq. (2). In an operator splitting method, the time evolution of $n(t)$ is still handled by the classical methods, such as the forward Euler method or Runge-Kutta method. Other than to integrate all aerosol dynamic processes together in one time step δ_t, operator splitting separates the integration into multiple fractional steps. There are various ways to construct the multiple fractional steps, such as $I_{\text{nuc}} \rightarrow G_{\text{cond}} \rightarrow C_{\text{coag}}$ or $\frac{1}{2}I_{\text{nuc}} \rightarrow G_{\text{cond}} \rightarrow \frac{1}{2}C_{\text{coag}}$, where 1/2 denotes a half integration step. The flowchart (**Figure 1**) shows an example how the operator splitting scheme $I_{\text{nuc}} \rightarrow G_{\text{cond}} \rightarrow C_{\text{coag}}$ is carried out. Details of the various models for the aerosol dynamics are described in the following subsections, and the error analyses for various operator splitting schemes are given in Section 3.

2.2.1. Nucleation

Nucleation is the gas-to-particle transition to generate nascent nanoparticle. There are various nucleation models, largely based on the classical nucleation theory developed in the 1920–1940s [34]. In the classical nucleation theory, there are two key parameters for the nucleation modeling, i.e., the nucleation rate (how many particles are nucleated per unit volume, unit time), and the critical size, which is the minimum size of nucleated particles with stable nucleus. According to the classical Becker-Döring theory [3, 35], the nucleation rate I and critical diameter d^* are

$$I = \frac{P_v x_v}{k_B T}\sqrt{\frac{2\sigma}{\pi m}} exp\left(-\frac{16\pi\sigma^3 m^2}{3(k_B T)^3 \rho_p^2 (\ln S)^2}\right), \tag{4}$$

$$d_p^* = \frac{4\sigma v_m}{k_B T \ln S}. \tag{5}$$

Here, P_v is the vapor pressure, x_v is the mole fraction of the vapor, m is the molecular mass of the nucleation vapor, ρ_p is the particle density, T is the temperature, $S = P_v/P_{\text{sat}}$ is the vapor saturation ratio, k_B is the Boltzmann constant, σ is the surface tension, and $v_m = m/\rho_p$ is the molecular effective volume of the nucleation vapor. Hence, the nucleation term in Eq. (2) can be expressed with the help of the Dirac delta function as

$$I_{\text{nuc}} = \int I\delta(d_p - d_p^*)\mathrm{d}d_p, \tag{6}$$

which means that only particles with the critical size are nucleated.

Within the OSMC, nucleation model is not limited to the classical nucleation theory. Any other nucleation model can be readily included, should only the nucleation rate and the critical size be provided.

2.2.2. Surface growth (condensation)

The functional form of the particle volume growth rate due to condensation depends on the Knudsen number $\text{Kn} = 2\lambda/d_p$, i.e., the ratio between the air mean free path λ and the particle radius $d_p/2$. In the free molecular regime (Kn ≥ 10) and the continuum regime (Kn ≤ 0.1), the condensational growth rates can be derived from the gas kinetic theory and the gradient diffusion model, respectively. In the transition regime (0.1 < Kn < 10), interpolation formulas are usually used, such as the harmonic mean of the free molecular and continuum regimes. The expressions for volume growth rate G_v and the condensation term G_{cond} are [3]

$$G_v = \begin{cases} \dfrac{\pi d_p^2 (P_v - P_p) v_m}{(2\pi m k_B T)^{1/2}}, & \text{Kn} \geq 10 \\ \dfrac{2\pi D d_p (P_v - P_p) v_m}{k_B T}, & \text{Kn} \leq 0.1 \\ harmonicmean, & 0.1 < \text{Kn} < 10 \end{cases} \tag{7}$$

$$G_{\text{cond}} = -\frac{\partial (n G_v)}{\partial v}. \tag{8}$$

Here, P_p is the equilibrium vapor pressure for particles of size d_p, which is usually assumed to be the saturation pressure of the condensation vapor. For particles close to the critical size, the Kelvin correction is also required [3].

Within the OSMC, any other surface growth model can also be accommodated, should only the particle volume growth rate be provided.

2.2.3. *Stochastic simulation of coagulation*

Coagulation is the process that two particles coalesce/aggregate to form a new particle. The coagulation dynamics is modeled by the well-known Smoluchowski's equation [3]

$$C_{\text{coag}} = \frac{1}{2}\int_0^v \beta(v, v-\tilde{v}) n(\tilde{v}) n(v-\tilde{v}) d\tilde{v} - \int_0^\infty \beta(v,\tilde{v}) n(v) n(\tilde{v}) d\tilde{v}. \tag{9}$$

The coagulation kernel function $\beta(v,u)$ describes the rate at which particles of size v coagulate with particles of size u. The first term on the right-hand side denotes the formation rate of particles with volume v due to coagulation between particles $\tilde{v}$ and $v - \tilde{v}$. (Volume is conserved before and after coagulation.) The half factor (1/2) is to correct the double-counting effect. The second term on the right-hand side describes the depletion rate of particles with volume v due to their coagulation with any other particles. Practically, the coagulation kernel function $\beta(v,u)$ is very hard to find, except for simplified conditions. In the free molecule regime (Kn ≥ 10), β has the form [3]

$$\beta(v,u) = \left(\frac{6}{\pi}\right)^{2/3}\left(\frac{\pi k_B T}{2\rho_p}\right)^{1/2}\left(\frac{1}{v}+\frac{1}{u}\right)^{1/2}(v^{1/3}+u^{1/3})^2. \tag{10}$$

In the continuum regime (Kn ≤ 0.1), β takes the form [3]

$$\beta(v,u) = \frac{2k_B T}{3\mu}\left(\frac{1}{v^{1/3}}+\frac{1}{u^{1/3}}\right)(v^{1/3}+u^{1/3}), \tag{11}$$

where μ is the dynamic viscosity of the gas-phase. In the transition regime (0.1 < Kn < 10), interpolation formula taking into account the former two regimes are generally used.

Although the Smoluchowski's equation was developed one century ago, and it has received broad attention in diverse fields, no analytical solutions are known for the kernels discussed above. Analytical solutions known so far are only limited to the following three simple kernels (or their linear combination) [5], i.e., (i) $\beta(v,u) = 1$, (ii) $\beta(v,u) = v + u$, and (iii) $\beta(v,u) = vu$.

Other than to solve the Smoluchowski's equation directly through various numerical means, the coagulation process is simulated by the Marcus-Lushnikov stochastic process in the OSMC. The Marcus-Lushnikov process is the natural stochastic analogue of the finite discrete Smoluchowski coagulation equation. The well-known review paper [5] provides a wide-ranging survey on the Marcus-Lushnikov process and the Smoluchowski's equation. Here, we follow the work of Gillespie [25], and give an intuitive introduction to the stochastic simulation of coagulation.

In a virtual box of volume V, there are N "well-mixed" particles. Any two particles may coalesce/aggregate to form a new one at a random time. The propensity C_{ij} of coalescence between two particles i and j is determined by the coagulation kernel $C_{ij} = \beta(i,j)/V$. Let $P(\tau,i,j)d\tau$ denote the probability that the next coalescence will occur in the time interval $(\tau,\tau + d\tau)$ between i and j $(i < j)$, then it is found [25]

$$P(\tau,i,j) = C_{ij}\exp\left[-\sum_{k=1}^{N-1}\sum_{l=k+1}^{N} C_{kl}\tau\right]. \tag{12}$$

In essence, the simulation is to randomly pick up two particles i and j to coalesce (forming a new particle) at the random time τ according to the joint probability density function $P(\tau,i,j)$. Repeat the former process as time advances. During the simulation, the number of particles drops, and the particle number density drops also.

The key step in the simulation is to generate a random point (τ,i,j) according to the density function $P(\tau,i,j)$. The random coagulation time τ satisfies a Poisson process (the exponential distribution), and it can be generated as

http://dx.doi.org/10.5772/65140

$$\tau = \frac{1}{C_0} \ln\left(\frac{1}{r_1}\right), \tag{13}$$

where $C_0 = \sum_{i=1}^{N-1} \sum_{j=i+1}^{N} C_{ij}$, and r_1 is a random number picked from the uniform distribution in the unit interval. Eq. (13) gives the classical way to generate a random variate from the exponential distribution with parameter C_0.

After selecting the time at which two particles coagulate, the next step is to randomly select two particles according to the probability function $P_2(i,j) = C_{ij}/C_0$. Loosely speaking, C_0 is the total coagulation probability between any two particles, and $P_2(i,j)$ is the fraction of the coagulation probability C_{ij} over C_0. Gillespie [25] introduced two methods to choose i and j, the so-called full conditioning and partial conditioning methods. In the full conditioning method, the selection of i and j is accomplished in two steps. During the first step, calculate the partial sum $C_i = \sum_{k=i+1}^{N} C_{ik}$, $(i = 1,...,N)$, then combine the N partial sums in sequential to form a range with the index i from 1 to N denoting the part for each partial sum, and then it is straightforward to build up a linear mapping between the range and the unit interval (0,1). Draw a random number from the uniform distribution, and check the random number landing on which corresponding part in the range, and then select particle i by the part index. During the second step, combine C_{ij}, $(j = i+1,...,N)$ in sequential to form another range, and build up the linear mapping between the new range with the unit interval, then draw another random number to determine the particle j in a similar way as in the first step. In the partial conditioning method, it is actually an acceptance-rejection method to generate a bivariate random number (i,j) from the function $P_2(i,j)$. In a general acceptance-rejection algorithm to generate a vector random variable $\vec{X}$ with density $f(\vec{x})$, it needs to choose a density function $g(\vec{x})$ and a constant $c \geq 1$ such that $cg(\vec{x}) \geq f(\vec{x})$. The algorithm proceeds as follows [36]:

1. Generate $\vec{Y}$ from density g.

2. Generate $\vec{U}$ from the uniform distribution on the interval $(0, cg(\vec{Y}))$.

3. If $\vec{U} \leq f(\vec{Y})$, then set $\vec{X} = \vec{Y}$; else, return to 1.

The density function $g(\vec{x})$ can be chosen freely. However, the optimal choice is to make the interval in 2 as small as possible, so that there is a better chance that $\vec{U} \leq f(\vec{Y})$ so as to accept $\vec{Y}$ as the required $\vec{X}$ without too often rejections and repetitions of the whole process. In the acceptance-rejection method [25] to generate i and j according to $P_2(i,j)$, $g(\vec{x})$ is chosen as a uniform function, and c is an upper bound of all $C_{i,j}$ (preferably the smallest upper bound). It is argued [25] that when the values of C_{ij} are relatively comparable with each other (ideally constant) the particle conditioning method should be more efficient than the full conditioning method.

Within the OSMC, both the full and particle conditioning methods have been implemented to choose from easily.

2.3. Full algorithm for the OSMC

Figure 1 shows the flowchart of OSMC including all aerosol dynamics. Only the first-order operator splitting is sketched in the figure.

At time $t = 0$, the simulator is initialized with given parameters, i.e., number of Monte Carlo (MC) particles N, simulator size V, and the initial PSD. The simulator size is determined as $V = N/M_0$, where M_0 is the initial number density. If simulation starts from an empty case, i.e., $M_0 = 0$, then the simulator is initialized with N particles of uniform size (the real value of the size is immaterial to the simulation results), and the simulator size is set to a huge value, say $V = 10^{10}$, which renders a tiny particle number density to approximate the condition $M_0 = 0$. If simulation starts from a case with a specific PSD, then the size of the N simulation particles is assigned by a stochastic process to satisfy the initial PSD. There are various convenient ways to generate random number to satisfy a given distribution [37], although they are usually not necessary in Monte Carlo simulation of aerosol dynamics, since mostly the simulation starts from an empty case.

In the particles nucleation step, the simulator size V is adjusted to reflect the change of particle number density due to nucleation. Nascent nucleated particles are added to the pool of MC particles. If the total number of MC particles exceeds the maximum allowable threshold value (which is set beforehand to balance the stochastic error and numerical cost), then down-sampling is performed, i.e., exceeding MC particles are randomly removed from the pool to satisfy the number constraint. And then every MC particle undergoes the surface growth process according to its growth rate (usually depends on its size).

The coagulation simulation process (those steps grouped in the dashed bounding box X_s in **Figure 1**) shows how to implement the stochastic algorithm of Gillespie [25]. Updating coagulation kernel is to calculate the pair coagulation rate $C_{ij} = \beta(v_i, v_j)/V$, $(i = 1,...,N - 1,$ $j = i + 1,...,N)$. The random coagulation time τ is generated according to Eq. (13). The comparison statement $\tau_{\text{sum}} + \tau < \delta t$ is to judge whether the time for two particles to coagulate is still admissible within the discrete time step δt. Here τ_{sum} is the accumulated coagulation time in the coagulation step, which is initialized to zero before a coagulation step begins. If the comparison statement is true, two particles are selected to perform coagulation, and the number of MC particles decreases by one. The subsequent up-sampling step is to keep the number of MC particles above a given threshold to avoid severe stochastic error.

3. Error analyses for operator splitting

Aerosol dynamics usually contain nucleation, surface growth, and coagulation processes. These processes have very different effects on the evolution of the total number density and volume fraction, which are the two most important statistics of aerosol particles. For the number density, it increases due to nucleation while decreases due to coagulation, and it does not change in the surface growth process (mainly condensation). For the volume fraction, it is conserved in the coagulation process. The volume fraction is generally dominated by surface growth process, and nucleation has limited contribution to the volume fraction directly, because nucleated particles are very small in size.

Nucleation is generally a function of the temperature and the nucleated vapor concentration, and it does not depend on the aerosol status (number density, particle size, etc). However, both condensation and coagulation are closely related to the number density, which is largely determined by the nucleation process. On the other hand, condensation and coagulation are usually loosely coupled through particle size and number density. Therefore, it is convenient and meaningful to consider two isolated cases: (i) nucleation + coagulation, which determine the number density, and (ii) nucleation + condensation, which determine the volume fraction.

3.1. Case of nucleation + coagulation

Before discussing the case of nucleation and coagulation, we introduce an integral transformation of the Smoluchowski's equation (9), which is very useful to understand the evolution of the moments of the PSD. Suppose $F(v)$ is an arbitrary function of v, and let

$$M(t) = \int_0^\infty F(v)n(v,t)\mathrm{d}v, \tag{14}$$

then from the Smoluchowski's equation (9), it can be derived [1]

$$\frac{\mathrm{d}M(t)}{\mathrm{d}t} = \frac{1}{2}\int_0^\infty\int_0^\infty [F(v+u) - F(v) - F(u)]\beta(v,u)n(v,t)n(u,t)\mathrm{d}u\mathrm{d}v. \tag{15}$$

Specifically, when $F(v) = v^k, (k \geq 0)$, $M_k(t)$ is the kth order moment, and Eq. (15) transforms to the dynamic equation for the moment $M_k(t)$

$$\frac{\mathrm{d}M_k(t)}{\mathrm{d}t} = \frac{1}{2}\int_0^\infty\int_0^\infty [(v+u)^k - v^k - u^k]\beta(v,u)n(v,t)n(u,t)\mathrm{d}u\mathrm{d}v. \tag{16}$$

One interesting but somewhat overlooked conclusion [1] is that the right hand side of Eq. (16) is negative if $k < 1$, and it is positive if $k > 1$, which mean that M_k will grow with time for

$k > 1$, and it will decrease with time for $k < 1$. Especially, $M_1(t)$ is constant with time. Actually, M_1 is the volume fraction, which is conserved during the pure coagulation process. M_0 is the total number density, which does decay due to coagulation. It is worth pointing out that although fractional moments (with k being a fraction number) do not possess obvious physical meaning, they are very useful in the method of moments with the interpolative closure [21] or with the Taylor expansion closure [22].

In the case of constant rate nucleation and constant kernel coagulation, the number density (M_0) satisfies the following simple model equation

$$\frac{\mathrm{d}M_0}{\mathrm{d}t} = J + KM_0^2. \tag{17}$$

One the right-hand side, the first term models the contribution to the number density from nucleation, and J represents the nucleation rate (number of particles nucleated per unit volume per unit time). The second term models the contribution from coagulation, and $K = -\frac{1}{2}\beta_0$, where β_0 is a constant coagulation kernel. In fact, in the moment dynamic Eq. (16) if the coagulation kernel is the constant β_0 and let $k = 0$ (to obtain the number density equation); we immediately have $-\frac{1}{2}\beta_0 M_0^2$ on the right-hand side. The notation $K = -\frac{1}{2}\beta_0$ (be aware of the negative sign) is found to be able to greatly simplify the following analyses.

The assumption of a constant nucleation rate and a constant coagulation kernel does not severely limit the applicability of the following discussion on the operator splitting errors for the more general case of nucleation and coagulation. The analyses focus on the local error in a generally small single time step. Hence, even a time-dependent nucleation rate $J(t)$ can be readily assumed constant within the single time step. On the other hand, the coagulation kernel β is a bivariate function of colliding particles sizes. Mathematically, it is possible to find a β_0 that satisfies

$$\beta_0 M_0^2 = \beta_0 \int_0^\infty \int_0^\infty n(v,t)n(u,t)\mathrm{d}u\mathrm{d}v = \int_0^\infty \int_0^\infty \beta(v,u)n(v,t)n(u,t)\mathrm{d}u\mathrm{d}v. \tag{18}$$

Of course, the right β_0 depends not only on the specific β function, but also on the distribution n. Here, we assume that a good estimate of β_0 is available and will not delve into this issue further.

Given the initial condition $M_0(t = 0) = M_0^*$, the model Eq. (17) has analytical solution

$$M_0(t) = \frac{\sqrt{JK}}{K}\tan\left[t\sqrt{JK} + \arctan\left(\frac{M_0^* K}{\sqrt{JK}}\right)\right]. \tag{19}$$

In fact, a solution of the same Eq. (17) is found as [7]

$$M_0(t) = \frac{\sqrt{-JK}}{K}\tanh\left[t\sqrt{-JK} + \operatorname{arctanh}\left(\frac{M_0^* K}{\sqrt{-JK}}\right)\right]. \tag{20}$$

Since, $K = -\frac{1}{2}\beta_0 < 0$ the solution Eq. (19) involves imaginary numbers. Using their relation to the complex exponential function, it is easy to show that Eqs. (19) and (20) are actually equivalent. Although the solution Eq. (20) (with K replaced by $-\frac{1}{2}\beta_0$) is more natural, the form in Eq. (19) proves more handy in the following analyses. Here, we expand Eq. (19) at $t = \delta_t$ as Taylor series for later usage

$$M_0(\delta_t) = M_0^* + (KM_0^{*2} + J)\delta_t + K(KM_0^{*2} + J)M_0^*\delta_t^2 + \frac{1}{3}K(3K^2M_0^{*4} + 4JKM_0^{*2} + J^2)\delta_t^3 + \mathcal{O}(\delta_t^4). \tag{21}$$

On the other hand, we can try to use the operator splitting method to solve the same Eq. (17), and compare the obtained solutions with the true solution (19) to quantify the operator splitting errors. There are various operator splitting schemes. Here, we will focus on two first-order Lie schemes and two second-order Strang schemes, i.e., (i) $I_{\text{nuc}} \to C_{\text{coag}}$, (ii) $C_{\text{coag}} \to I_{\text{nuc}}$, (iii) $\frac{1}{2}I_{\text{nuc}} \to G_{\text{coag}} \to \frac{1}{2}I_{\text{nuc}}$, and (iv) $\frac{1}{2}C_{\text{coag}} \to I_{\text{nuc}} \to \frac{1}{2}C_{\text{coag}}$.

In the operator splitting method, the following two joint equations are solved

$$\frac{dM_0^{(1)}}{dt} = J, \tag{22a}$$

$$\frac{dM_0^{(2)}}{dt} = KM_0^2. \tag{22b}$$

(i) Lie scheme $I_{\text{nuc}} \to C_{\text{coag}}$ (Denoted as Lie1)

To seek the solution at $t = \delta_t$, Eq. (22a) is solved first with the given initial condition at $t = 0$, then Eq. (22b) is solved with the just obtained solution at $t = \delta_t$ as the initial condition.

Eq. (22a) has solution:

$$M_0^{(1)}(t) = Jt + M_0^*. \tag{23}$$

Hence, $M_0^{(1)}(\delta_t) = J\delta_t + M_0^*$. Take $M_0^{(1)}(\delta_t)$ as the initial condition and solve Eq. (22b), then the final solution is found as

$$M_0^{(OP)}(t) = -\frac{J\delta_t + M_0^*}{JK\delta_t t + KM_0^* t - 1}. \tag{24}$$

Let $t = \delta_t$ in the above equation, and expand it into Taylor series

$$M_0^{(OP)}(\delta_t) = M_0^* + (KM_0^2 + J)\delta_t + K(KM_0^{*2} + 2J)M_0^*\delta_t^2 + \mathcal{O}(\delta_t^3). \tag{25}$$

The operator splitting error can be found by comparing the two Taylor expansion series (25) and (21)

$$Er_{0i)} = M_0^{(OP)}(\delta_t) - M_0(\delta_t) = M_0^* JK\delta_t^2 + \mathcal{O}(\delta_t^3) = -\frac{1}{2}M_0^* J\beta_0\delta_t^2 + \mathcal{O}(\delta_t^3). \tag{26}$$

(ii) Lie scheme $I_{\text{coag}} \rightarrow C_{\text{nuc}}$ (Denoted as Lie2)

The analysis is quite similar to case (i), except that the solution procedure for the two joint Eqs. (22a) and (22b) is reversed. The final solution is

$$M_0^{(OP)}(\delta_t) = J\delta_t - \frac{M_0^*}{KM_0^*\delta_t - 1}, \tag{27}$$

and its Taylor expansion is

$$M_0^{(OP)}(\delta_t) = M_0^* + (KM_0^2 + J)\delta_t + K^2 M_0^{*3}\delta_t^2 + \mathcal{O}(\delta_t^3). \tag{28}$$

Hence, the operator splitting error is

$$Er_{0ii)} = -M_0^* JK\delta_t^2 + \mathcal{O}(\delta_t^3) = \frac{1}{2}M_0^* J\beta_0\delta_t^2 + \mathcal{O}(\delta_t^3). \tag{29}$$

(iii) Strang scheme $\frac{1}{2}I_{\text{nuc}} \rightarrow C_{\text{coag}} \rightarrow \frac{1}{2}I_{\text{nuc}}$ (Denoted as Strang1)

Here, the solution to the joint Eqs. (22) is carried out in three steps. First, Eq. (22a) is solved with the initial condition $M_0(0) = M_0^*$ to get the solution $M_0^{(1)}(t)$. Second, $M_0^{(1)}(\frac{1}{2}\delta_t)$ is used as the initial condition to solve Eq. (22b) to get the solution $M_0^{(2)}(t)$. Finally, $M_0^{(2)}(\delta_t)$ is used as the initial condition at $t = \frac{1}{2}\delta_t$ to solve Eq. (22a) again. The final solution at $t = \delta_t$ is found as

$$M_0^{(\mathrm{OP})}(\delta_t) = \frac{1}{2} J\delta_t - \frac{J\delta_t + 2M_0^*}{JK\delta_t^2 + 2KM_0^*\delta_t - 2}, \tag{30}$$

and its Taylor expansion is

$$M_0^{(\mathrm{OP})}(\delta_t) = M_0^* + (KM_0^{*2} + J)\delta_t + K(KM_0^{*2} + J)M_0^*\delta_t^2 + \frac{1}{4}K(4K^2M_0^{*4} + 6KJM_0^{*2} + J^2)\delta_t^3 + \mathcal{O}(\delta_t^4). \tag{31}$$

Hence, the operator splitting error is

$$Er_{0\mathrm{iii})} = \frac{JK}{12}(2M_0^{*2}K - J)\delta_t^3 + \mathcal{O}(\delta_t^4) = \frac{J\beta_0}{24}(M_0^{*2}\beta_0 + J)\delta_t^3 + \mathcal{O}(\delta_t^4). \tag{32}$$

(iv) Strang scheme $\frac{1}{2}C_{\mathrm{coag}} \to I_{\mathrm{nuc}} \to \frac{1}{2}C_{\mathrm{coag}}$ (Denoted as Strang2)

Similar to case (iii), the final solution is found as

$$M_0^{(\mathrm{OP})}(\delta_t) = 2\frac{JKM_0^*\delta_t^2 - 2J\delta_t - 2M_0^*}{-JK^2M_0^*\delta_t^3 + 2JK\delta_t^2 + 4KM_0^*\delta_t - 4}, \tag{33}$$

and its Taylor expansion is

$$M_0^{(\mathrm{OP})}(\delta_t) = M_0^* + (KM_0^{*2} + J)\delta_t + K(KM_0^{*2} + J)M_0^*\delta_t^2 + \frac{1}{4}K(4K^2M_0^{*4} + 5KJM_0^{*2} + 2J^2)\delta_t^3 + \mathcal{O}(\delta_t^4). \tag{34}$$

Hence, the operator splitting error is

$$Er_{0\mathrm{iv})} = \frac{JK}{12}(-M_0^{*2}K + 2J)\delta_t^3 + \mathcal{O}(\delta_t^4) = -\frac{J\beta_0}{48}(M_0^{*2}\beta_0 + 4J)\delta_t^3 + \mathcal{O}(\delta_t^4). \tag{35}$$

To summarize, the two Lie schemes have the same magnitude in splitting error, both are of second order in δ_t within a time step. So the global integration error during time t_{all} is of order δ_t, since the error will accumulate in t_{all}/δ_t steps. The Lie1 scheme underestimates the number density, while the Lie2 scheme overestimates that. This is because coagulation rate closely depends on the number density, the higher the number density, the faster the coagulation rate. Coagulation rate would be overpredicted if nucleation is integrated before coagulation, hence the number density would be underpredicted as in Lie1, and vice versa.

The splitting errors in the two Strang schemes are of third order in δ_t within a time step, which renders them a second-order scheme globally. The Strang1 scheme is found to overestimate the number density, while the two Strang2 scheme to underestimate that. The magnitude of the absolute errors in these two schemes can be compared through investigating the coefficients of δ_t^3 in Eqs. (32) and (35)

$$2(M_0^{*2}\beta_0 + J) - (M_0^{*2}\beta_0 + 4J) = M_0^{*2}\beta_0 - 2J. \tag{36}$$

If $\frac{1}{2}M_0^{*2}\beta_0 < J$, i.e., coagulation rate is smaller than nucleation rate, the absolute splitting error in Eq. (32) is smaller than that in Eq. (35). When $\frac{1}{2}M_0^{*2}\beta_0 \ll J$, the error in Eq. (33) approaches a half of that in Eq. (36). On the other hand, when $\frac{1}{2}M_0^{*2}\beta_0 \gg J$, the error in Eq. (32) nearly doubles that in Eq. (35). Hence, in the Strang schemes it is preferable to split the higher rate process into symmetric substeps to reduce the splitting error.

Only the most common first and second-order operator splitting schemes are discussed so far. They are systematical ways to construct even higher order schemes. Especially, a third-order scheme [38] can be obtained through the combination of the two Lie and two Strang schemes. Here, we are not interested to discuss the even higher order schemes other than the Lie and Strang schemes, since higher order schemes are complicate to implement and may not be worthy for the solving the aerosol dynamics. One interesting conclusion [38] is that the scheme $\frac{2}{3}(\text{Strang1} + \text{Strang2}) + \frac{1}{4}(\text{Lie1} + \text{Lie2})$ is of third order, which can be used to check the accuracy of the above derivations. Keep the third-order term δ_t^3 in the error functions (26), (29), (32), and (29), then calculate the error in the third-order scheme constructed as mentioned shortly before, it is found that only terms involving δ_t^4 and higher orders survive. This is an indirect proof that all the derivations of the error functions are correct.

Following, we will briefly discuss the splitting errors for higher order moments other than M_0. Since the total volume is conserved during coagulation, nucleation, and coagulation are independent of each other with respect to M_1 (volume fraction). Hence, there are no splitting errors for predicting M_1 for M_2, the model equation for nucleation and coagulation can be easily derived from Eqs. (6) and (16) as

$$\frac{dM_2}{dt} = Jv^{*2} + \frac{1}{2}\beta_0\int_0^\infty\int_0^\infty (2vu)n(v,t)n(u,t)\mathrm{d}u\mathrm{d}v = Jv^{*2} + M_1^2, \tag{37}$$

and M_1 is determined by nucleation only, which can be integrated as

$$M_1(t) = Jv^* t + M_1^*. \tag{38}$$

Here, v^* is the critical volume of nucleated particle, and M_1^* is the initial value at $t = 0$. The solution to Eq. (37) is

$$M_2(t) = \frac{1}{3}J^2 v^{*2}\beta_0 t^3 + Jv^* M_1^*\beta_0 t^2 + (Jv^{*2} + \beta_0 M_1^{*2})t + M_2^*. \tag{39}$$

Similar to the above error analyses, the four splitting errors for solving Eq. (37) can be readily obtained as

$$Er_{2(\mathrm{i})} = \frac{1}{3}\beta_0 Jv^*(2Jv^*\delta_t + 3M_1^*)\delta_t^2, \tag{40a}$$

$$Er_{2(\mathrm{ii})} = -\frac{1}{3}\beta_0 Jv^*(Jv^*\delta_t + 3M_1^*)\delta_t^2, \tag{40b}$$

$$Er_{2(\mathrm{iii})} = -\frac{1}{12}\beta_0 J^2 v^{*2}\delta_t^3, \tag{40c}$$

$$Er_{2(\mathrm{iv})} = \frac{1}{6}\beta_0 J^2 v^{*2}\delta_t^3. \tag{40d}$$

The splitting errors for the two second-order Strang schemes are independent of the initial conditions.

3.2. Case of nucleation + condensation

Here, a case with constant nucleation rate and constant condensation growth in diameter is considered. A constant growth rate in diameter is not arbitrary at all. Eq. (7) gives the volume growth rate due to condensation as a piecewise function. The particle diameter growth rate function can be obtained easily through the sphere volume formula $v = \frac{\pi}{6}d_p^3$. The detailed function can also be found in reference [31]. In fact, the particle diameter growth rate in the free molecular regime is independent of its size, which can be seen as a constant determined by environmental parameters (vapor pressure and temperature). In the following discussion, the particle diameter growth rate is assumed as G. In this subsection, the moments are defined with respect to the particle diameter d_p other than the volume $M_k = \int n(d_p) d_p^k \mathrm{d}d_p$. No attempt is sought to distinguish the symbols for moments defined differently, with the hope to reduce new symbols and to bring no confusion.

The dynamic equations for moments due to pure nucleation (to generate particles with diameter d_p^* at rate J) can be easily obtained with the help of the Delta function (6), which are

$$\frac{\mathrm{d}M_0^{(1)}}{\mathrm{d}t} = J, \tag{41a}$$

$$\frac{\mathrm{d}M_1^{(1)}}{\mathrm{d}t} = Jd_p^*, \tag{41b}$$

$$\frac{\mathrm{d}M_2^{(1)}}{\mathrm{d}t} = Jd_p^{*2}, \tag{41c}$$

$$\frac{\mathrm{d}M_3^{(1)}}{\mathrm{d}t} = Jd_p^{*3}. \tag{41d}$$

The dynamic equations for moments due to condensation (at a constant diameter growth rate G) are

$$\frac{\mathrm{d}M_0^{(2)}}{\mathrm{d}t} = \int -\frac{\partial(nG)}{\partial d_p} \mathrm{d}d_p = 0, \tag{42a}$$

$$\frac{\mathrm{d}M_1^{(2)}}{\mathrm{d}t} = \int -\frac{\partial(nG)}{\partial d_p} d_p \mathrm{d}d_p = G\int n \mathrm{d}d_p = GM_0^{(2)}, \tag{42b}$$

$$\frac{\mathrm{d}M_2^{(2)}}{\mathrm{d}t} = \int -\frac{\partial(nG)}{\partial d_p} d_p^2 \mathrm{d}d_p = 2G\int n d_p \mathrm{d}d_p = 2GM_1^{(2)}, \tag{42c}$$

$$\frac{\mathrm{d}M_3^{(2)}}{\mathrm{d}t} = \int -\frac{\partial(nG)}{\partial d_p} d_p^3 \mathrm{d}d_p = 3G\int n d_p^2 \mathrm{d}d_p = 3GM_2^{(2)}. \tag{42d}$$

Hence, the dynamic moments equations for nucleation and condensation are

$$\frac{\mathrm{d}M_0}{\mathrm{d}t} = J, \tag{43a}$$

$$\frac{\mathrm{d}M_1}{\mathrm{d}t} = Jd_p^* + GM_0, \tag{43b}$$

$$\frac{\mathrm{d}M_2}{\mathrm{d}t} = Jd_p^{*2} + 2GM_1, \tag{43c}$$

$$\frac{\mathrm{d}M_3}{\mathrm{d}t} = Jd_p^{*3} + 3GM_2. \tag{43d}$$

Assume the initial condition at $t = 0$ is $[M_0^*, M_1^*, M_2^*, M_3^*]$, then Eq. (43) has solution

$$M_0(t) = Jt + M_0^*, \tag{44a}$$

$$M_1(t) = \frac{1}{2}JGt^2 + (Jd_p^* + GM_0^*)t + M_1^*, \tag{44b}$$

$$M_2(t) = \frac{1}{3}JG^2t^3 + (Jd_p^* + GM_0^*)Gt^2 + (Jd_p^{*2} + 2GM_1^*)t + M_2^*, \tag{44c}$$

$$M_3(t) = \frac{1}{4}JG^3t^4 + (Jd_p^* + GM_0^*)G^2t^3 + \frac{3}{2}(Jd_p^{*2} + 2GM_1^*)Gt^2 + (Jd_p^{*3} + 3GM_2^*)t + M_3^*. \tag{44d}$$

On the other hand, the operator splitting can be used to solve Eqs. (41) and (42). Parallel to the analyses in Section 3.1, four splitting schemes are considered as follows.

(i) Lie scheme $I_{\mathrm{nuc}} \rightarrow G_{\mathrm{cond}}$ (Denoted as Lie1)

After solving Eqs. (41) and (42) sequentially, the final solution at $t = \delta_t$ is found as

$$M_0^{\mathrm{OP}}(t) = Jt + M_0^*, \tag{45a}$$

$$M_1^{\mathrm{OP}}(t) = JGt^2 + (Jd_p^* + GM_0^*)t + M_1^*, \tag{45b}$$

$$M_2^{\mathrm{OP}}(t) = JG^2t^3 + (2Jd_p^* + GM_0^*)Gt^2 + (Jd_p^{*2} + 2GM_1^*)t + M_2^*, \tag{45c}$$

$$M_3^{\mathrm{OP}}(t) = JG^3t^4 + (3Jd_p^* + GM_0^*)G^2t^3 + (3Jd_p^{*2} + 3GM_1^*)Gt^2 + (Jd_p^{*3} + 3GM_2^*)t + M_3^*. \tag{45d}$$

The splitting error is found by comparing the splitting solution (45) with the exact solution (44)

$$Er_{0(\mathrm{i})} = 0, \tag{46a}$$

$$Er_{1(\mathrm{i})} = \frac{1}{2}JG\delta_t^2, \tag{46b}$$

$$Er_{2(\mathrm{i})} = Jd_p^*G\delta_t^2 + \frac{2}{3}JG^2\delta_t^3, \tag{46c}$$

$$Er_{3(\mathrm{i})} = \frac{3}{2}Jd_p^{*2}G\delta_t^2 + 2JG^2d_p^*\delta_t^3 + \frac{3}{4}JG^3\delta_t^4. \tag{46d}$$

(ii) Lie scheme $G_{\mathrm{cond}} \rightarrow I_{\mathrm{nuc}}$ (Denoted as Lie2)

After solving Eqs. (42) and (41) sequentially, the final solution at $t = \delta_t$ is found as

$$M_0^{\mathrm{OP}}(t) = Jt + M_0^*, \tag{47a}$$

$$M_1^{\mathrm{OP}}(t) = (Jd_p^* + GM_0^*)t + M_1^*, \tag{47b}$$

$$M_2^{\mathrm{OP}}(t) = G^2M_0^*t^2 + (Jd_p^{*2} + 2GM_1^*)t + M_2^*, \tag{47c}$$

$$M_3^{\mathrm{OP}}(t) = G^3M_0^*t^3 + 3G^2M_1^*t^2 + (Jd_p^{*3} + 3GM_2^*)t + M_3^*. \tag{47d}$$

http://dx.doi.org/10.5772/65140

The splitting error is found by comparing the splitting solution (47) with the exact solution (44)

$$Er_{0(\text{ii})} = 0, \tag{48a}$$

$$Er_{1(\text{ii})} = -\frac{1}{2} JG\delta_t^2, \tag{48b}$$

$$Er_{2(\text{ii})} = -Jd_p^* G\delta_t^2 - \frac{1}{3} JG^2\delta_t^3, \tag{48c}$$

$$Er_{3(\text{ii})} = -\frac{3}{2} Jd_p^{*2} G\delta_t^2 - 2Jd_p^* G^2\delta_t^3 - \frac{1}{4} JG^3\delta_t^4. \tag{48d}$$

(iii) Strang scheme $\frac{1}{2} I_{\text{nuc}} \rightarrow G_{\text{cond}} \rightarrow \frac{1}{2} I_{\text{nuc}}$ (Denoted as Strang1)

After sequentially solving Eq. (41) with half time step $\delta_t/2$, then Eq. (42) with full time step, and then Eq. (41) with half step again, the final solution at $t = \delta_t$ is found as

$$M_0^{\text{OP}}(t) = Jt + M_0^*, \tag{49a}$$

$$M_1^{\text{OP}}(t) = \frac{1}{2} JGt^2 + (Jd_p^* + GM_0^*)t + M_1^*, \tag{49b}$$

$$M_2^{\text{OP}}(t) = \frac{1}{2} JG^2t^3 + (Jd_p^* + GM_0^*)Gt^2 + (Jd_p^{*2} + 2GM_1^*)t + M_2^*, \tag{49c}$$

$$M_3^{\text{OP}}(t) = \frac{1}{2} JG^3t^4 + (\frac{3}{2} Jd_p^* + GM_0^*)G^2t^3 + (\frac{3}{2} Jd_p^{*2} + 3GM_1^*)Gt^2 + (Jd_p^{*3} + 3GM_2^*)t + M_3^*. \tag{49d}$$

The splitting error is found by comparing the splitting solution (49) with the exact solution (44)

$$Er_{0(\text{iii})} = 0, \tag{50a}$$

$$Er_{1(\text{iii})} = 0, \tag{50b}$$

$$Er_{2(\text{iii})} = \frac{1}{6} JG^2\delta_t^3, \tag{50c}$$

$$Er_{3(\mathrm{iii})} = \frac{1}{2} JG^2 d_p^* \delta_t^3 + \frac{1}{4} JG^3 \delta_t^4. \tag{50d}$$

(iv) Strang scheme $\frac{1}{2}G_{\mathrm{cond}} \to I_{\mathrm{nuc}} \to \frac{1}{2}G_{\mathrm{cond}}$ (Denoted as Strang2)

After sequentially solving Eq. (42) with half time step $\delta_t/2$, then Eq. (41) with full time step, and then Eq. (42) with half step again, the final solution at $t = \delta_t$ is found as

$$M_0^{\mathrm{OP}}(t) = Jt + M_0^*, \tag{51a}$$

$$M_1^{\mathrm{OP}}(t) = \frac{1}{2} JGt^2 + (Jd_p^* + GM_0^*)t + M_1^*, \tag{51b}$$

$$M_2^{\mathrm{OP}}(t) = \frac{1}{4} JG^2 t^3 + (Jd_p^* + GM_0^*)Gt^2 + (Jd_p^{*2} + 2GM_1^*)t + M_2^*, \tag{51c}$$

$$M_3^{\mathrm{OP}}(t) = \frac{1}{8} JG^3 t^4 + (\frac{3}{4} Jd_p^* + GM_0^*)G^2 t^3 + (\frac{3}{2} Jd_p^{*2} + 3GM_1^*)Gt^2 + (Jd_p^{*3} + 3GM_2^*)t + M_3^*. \tag{51d}$$

The splitting error is found by comparing the splitting solution (51) with the exact solution (44)

$$Er_{0(\mathrm{iv})} = 0, \tag{52a}$$

$$Er_{1(\mathrm{iv})} = 0, \tag{52b}$$

$$Er_{2(\mathrm{iv})} = -\frac{1}{12} JG^2 \delta_t^3, \tag{52c}$$

$$Er_{3(\mathrm{iv})} = -\frac{1}{4} JG^2 d_p^* \delta_t^3 - \frac{1}{8} JG^3 \delta_t^4. \tag{52d}$$

To summarize, there is no splitting error for the number density (M_0), which is solely determined by nucleation. For M_1 to M_3, all Lie schemes are of first order (globally), and the leading errors are of opposite sign between Lie1 and Lie2. For Strang schemes, the splitting error for M_1 is also zero, and the errors for M_2 and M_3 are of second order. In Lie1 and Strang1 scheme, the splitting solutions overpredict the high order moments, since nucleation in the prestep produces more than real particles to grow in condensation. On the other hand, Lie2 and Strang2 underpredict the high order moments. Since the splitting errors are in closed form of the time step δ_t, it is possible to construct higher order splitting scheme with zero splitting error for all moments.

Furthermore, the leading order (with respect to δ_t) splitting errors for any two subsequent moments differ by a factor of d_p^*, which renders the relative errors (splitting errors divided by the corresponding moments values) comparable among various moments, since two subsequent moments also differ by a factor of the particles average diameter according to the moment definition. The splitting errors are independent of the initial conditions, i.e., M_0^*, M_1^*, M_2^*, and M_3^*, which seems strange at first glance. In fact, the splitting errors are anchored to the moment's evolution. Take the M_1 splitting error in the Lie1 scheme as example, the error contains the factor JG, while the leading order in the analytical solution for M_1 in Eq. (58) also contains exactly the same factor JG. Then only the time step δ_t matters when calculating the relative error. Generally speaking, the relative errors due to the operator splitting are small, almost independent of the nucleation rate and the condensational growth rate.

4. Numerical examples

4.1. Case of nucleation + coagulation

Corresponding to the error analyses in Section 3.1, the case of constant nucleation and constant coagulation are simulated with the OSMC. The coagulation kernel and nucleation rate are set as unit $\beta_0 = 1$, and $J = 1$. Initially particles are mono dispersed with unit volume size, and the number density is $M_0^* = 1$. Nucleated particles also have unit volume size.

Figure 2 shows the number density (M_0) simulated by the four schemes, compared against the analytical solution (19). All the simulations use fixed time step $\delta_t = 0.1$, with $N = 2000$ Monte Carlo particles. The OSMC results are averaged over dozens of independent simulations to eliminate the randomness. Lie1 scheme is found to systematically underpredict the number density, while Lie2 to overpredict. Strang1 gives result negligibly higher than the analytical solution. Strang2 slightly underpredicts the true solution. All these findings agree extremely well with the analyses in Section 3.1. It is also found that Strang1 shows better precision than Strang2, that is because nucleation rate is higher than the coagulation rate ($J = 1 > \frac{1}{2}\beta_0 M_0^* = \frac{1}{2}$), and the higher rate process nucleation is split in Strang1, as explained in Section 3.1 also.

Figure 3 compares M_2 among the four splitting schemes and the analytical solution (39). Lie1 is found to overestimate M_2 a little bit, while Lie2 to underestimate. The second-order Strang schemes give result almost indistinguishable from the analytical solution. Although second-order Strang schemes give better results than the Lie schemes, the discrepancy is not so significant as that for M_0 in **Figure 2**.

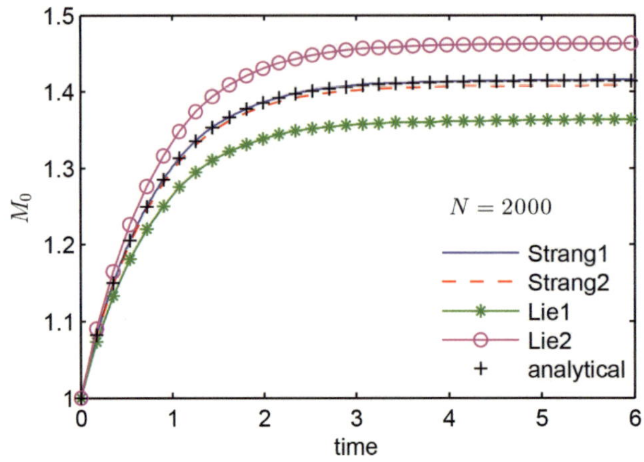

Figure 2. Evolution of total particle number density M_0 for constant kernel coagulation and constant rate nucleation ($\delta_t = 0.1$).

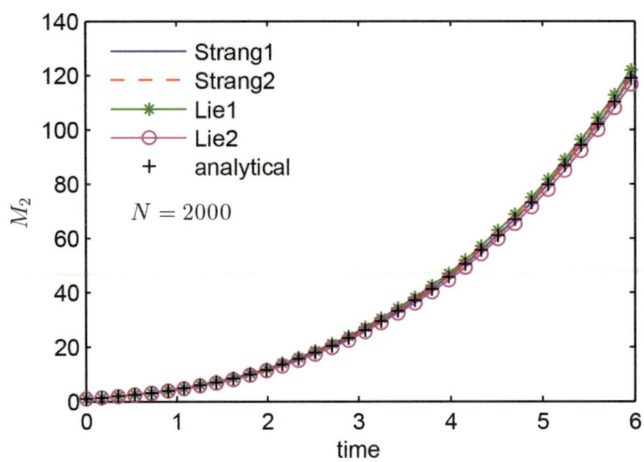

Figure 3. Evolution of moment M_2 in case of constant kernel coagulation and constant rate nucleation (moment is defined with respect to the particle volume).

Both the error analyses and the numerical examples above show that the second-order Strang schemes have much higher precision than the first-order Lie schemes. Within the OSMC, coagulation is the most computationally intensive part, and nucleation only makes a tiny fraction of the total time cost. The Strang1 scheme has much higher precision than the

Lie schemes, and only needs to integrate nucleation twice often as in the Lie schemes, where the computational overhead is negligible. Strang2 needs nearly double computational cost as Strang1, and Strang2 outperforms Strang1 in precision only if $\frac{1}{2}\beta_0 M_0^{*2} > J$. The best performance of Strang2 over Strang1 is to nearly double the precision when $\beta_0 M_0^{*2} \gg J$. Hence, Strang1 is the optimal splitting scheme taking both the numerical precision and cost into account.

Another general conclusion is that only M_0 needs to be concerned when considering the operator splitting errors. Operator splitting generally produces less errors for higher order moments. The M_2 splitting errors for the Strang schemes contain the factor of the critical volume of nucleated particles, which is generally very small. Taking a more practical numerical example of dibutyl phthalate (DBP) aerosol nucleation and coagulation in a turbulent mixing layer [26], the critical volume v^* is of order $10^{-27}\mathrm{m}^3$, which renders very small splitting errors for high order moments. However, the splitting error for M_0 may be still significant, and needs proper splitting scheme combined with small enough time step to satisfy the given error criteria.

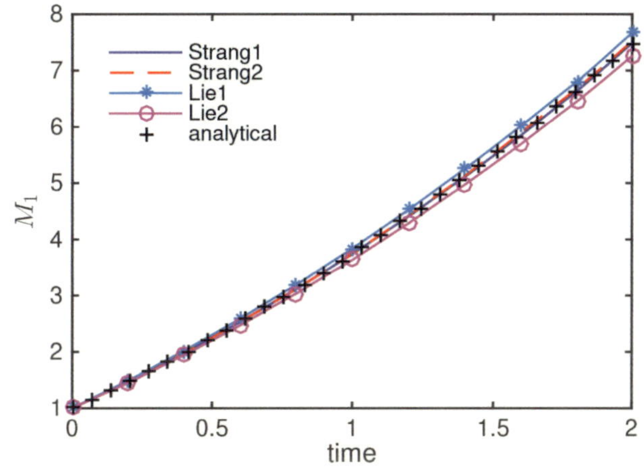

Figure 4. Evolution of moment M_1 in case of constant nucleation and constant condensational growth in diameter (moment is defined with respect to particle diameter).

4.2. Case of nucleation + condensation

Corresponding to the error analysis in Section 3.2, the case of constant nucleation and constant condensational growth in diameter is simulated with the OSMC. Both the nucleation rate and growth rate are set to unit $J = 1$ and $G = 1$. Initially particles are monodispersed with unit volume size, and the number density is $M_0^* = 1$. Nucleated particles also have unit volume size. The time step is $\delta_t = 0.2$.

Figures 4 and **5** show the evolution of moments M_1 and M_3, respectively, obtained by the four splitting schemes and the analytical solutions (44d). While M_3 directly represents the particle volume fraction (lacking a factor of $\pi/6$), M_1 do not have simple physical meaning, which may be used to define an average diameter through M_1/M_0. For both M_1 and M_3, the characteristics of splitting errors are quite similar. Lie1 is found to overpredict, while Lie2 to underpredict. The splitting errors for the two Strang schemes are very small, almost indistinguishable from the analytical solutions. According to the error analyses (49c) and (51c), the splitting errors for M_1 Strang schemes are zero. Close scrutiny shows in **Figure 4** that the numerical error is not actually zero, which actually arises from the ordinary differential equation (ODE) integration error, since a relatively large time step $\delta_t = 0.2$ is used. For a more appropriate smaller time step, both the ODE integration error and splitting error are much smaller. Comparing **Figures 4** and **5**, the relatively errors are comparable for both M_1 and M_3 here. All the findings agree very well with the analyses in Section 3.2.

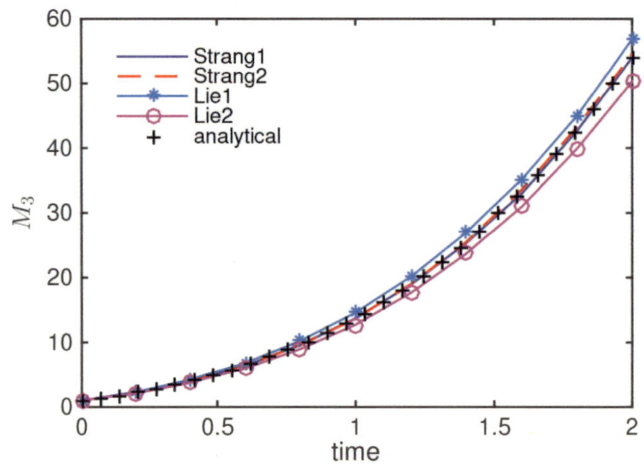

Figure 5. Evolution of moment M_3 (volume fraction missing a factor $\pi/6$) in case of constant nucleation and constant condensational growth in diameter (moment is defined with respect to particle diameter).

5. Conclusions

The operator splitting Monte Carlo (OSMC) method has been developed recently [26] for solving the population balance equation for aerosol dynamics. Within the OSMC, nucleation and surface growth are handled with deterministic means, while coagulation is simulated with a stochastic method (the Marcus-Lushnikov stochastic process). The deterministic and stochastic approaches in the algorithm are synthesized under the framework of operator splitting. The ultimate goal of the OSMC is to greatly improve the numerical efficiency and to preserve the extraordinary flexibility and applicability of a stochastic method for aerosol dynamics simulation.

The OSMC has been validated through quite a few testing cases [26], and has also been used to simulate aerosol evolution under various conditions [32, 39]. However, the operator splitting errors in the OSMC have not been systematically investigated. Here, focused on two representative cases, i.e., constant nucleation and coagulation, and constant nucleation and condensation, the splitting errors for four splitting schemes (two Lie schemes and two Strang schemes) are analyzed rigorously, combined with concrete numerical examples.

The number density, one of the most important statistics for aerosol particles, is found to be underestimated for the Lie1 scheme ($I_{\text{nuc}} \rightarrow C_{\text{coag}}$), while overestimated for the Lie2 scheme ($C_{\text{coag}} \rightarrow I_{\text{nuc}}$). This is because coagulation rate would be overpredicted if nucleation is integrated before coagulation, hence the number density would be underpredicted as in Lie1, and vice versa. The second-order Strang schemes exhibit much better precision that the first-order Lie schemes. If coagulation rate is higher than the nucleation rate, Strang2 ($\frac{1}{2}C_{\text{coag}} \rightarrow I_{\text{nuc}} \rightarrow \frac{1}{2}C_{\text{coag}}$) has better precision than Strang1 ($\frac{1}{2}I_{\text{nuc}} \rightarrow G_{\text{coag}} \rightarrow \frac{1}{2}I_{\text{nuc}}$). Otherwise, Strang1 has better precision. However, for the OSMC, Strang1 is always the most preferable splitting scheme after taking the numerical cost and precision into account.

The operator splitting errors for the case of nucleation and condensation show that splitting errors for different moments (except for M_0) are comparable with each other. Splitting between nucleation and condensation has marginal impact on accurately predicting the moments. This is in great contrast to the prediction of M_0, where splitting between nucleation and coagulation has significant impact.

To construct a good operator splitting scheme for the general case of nucleation, coagulation, and condensation the first priority is to consider the splitting error for M_0. When a good scheme for splitting between nucleation and coagulation is constructed, splitting among other aerosol dynamic processes can be constructed at more freedom. The overall splitting errors are pretty much determined by the splitting error for M_0.

The analyses not only provide sound theoretical bases for selecting the most efficient operator splitting scheme for the usage of the OSMC, but also shed lights on how to adopt operator splitting in other PBE solving methods, i.e., direct discretization, method of moments etc.

Acknowledgements

The financial support from the National Natural Science Foundation of China (No. 11402179 and 11572274) is greatly acknowledged. KZ would like to thank Dr. F. Bisetti at KAUST, Saudi Arabia, for his insightful discussion and valuable help. The financial support from the General Research Fund, Research Grants Council of the Hong Kong SAR, China (No.

PolyU 152663/16E) is also greatly appreciated for allowing the authors' further development and extension of this OSCM foundation method.

Author details

Kun Zhou[1*] and Tat Leung Chan[2]

*Address all correspondence to: zhou.kun@wust.edu.cn

1 The State Key Laboratory of Refractories and Metallurgy, Wuhan University of Science and Technology, Wuhan, China

2 Department of Mechanical Engineering, The Hong Kong Polytechnic University, Kowloon, Hong Kong

References

[1] R. L. Drake. A general mathematical survey of the coagulation equation. In G. M. Hidy and J. R. Brock, editors, *Topics in Current Aerosol Research,*. Pergamon Press, New York, pp. 201–376, 1972.

[2] D. Ramkrishna. *Population Balances: Theory and Applications to Particulate Systems in Engineering*. Academic Press, San Diego, 2000.

[3] S. K. Friedlander. *Smoke, Dust, and Haze: Fundamentals of Aerosol Dynamics*. Oxford University Press, New York, second edition, 2000.

[4] M. Smoluchowski. Drei Vorträge über Diffusion, Brownsche Bewegung und Koagulation von Kolloidteilchen. *Physikalische Zeitschrift*, 17:557–585, 1916.

[5] D. J. Aldous. Deterministic and stochastic models for coalescence (aggregation and coagulation): a review of the mean-field theory for probabilists. *Bernoulli*, 5:3–48, 1999.

[6] T. E. Ramabhadran, T. W. Peterson, and J. H. Seinfeld. Dynamics of aerosol coagulation and condensation. *AIChE Journal*, 220(5):840–851, 1976.

[7] S. C. Davies, J. R. King, and J. A. D. Wattis. The Smoluchowski coagulation equations with continuous injection. *Journal of Physics A: Mathematical and Theoretical*, 32:7745–7763, 1999.

[8] F. Gelbard, Y. Tambour, and J. H. Seinfeld. Sectional representations for simulating aerosol dynamics. *Journal of Colloid and Interface Science*, 760(2):541–556, 1980.

[9] S. Rigopoulos and A. G. Jones. Finite-element scheme for solution of the dynamic population balance equation. *AIChE Journal*, 490(5):1127–1139, 2003.

[10] A. H. Alexopoulos, A. I. Roussos, and C. Kiparissides. Part I: Dynamic evolution of the particle size distribution in particulate processes undergoing combined particle growth and aggregation. *Chemical Engineering Science*, 59(24):5751–5769, 2004.

[11] S. Ganesan. An operator-splitting Galerkin/SUPG finite element method for population balance equations : stability and convergence. *ESAIM. Mathematical Modelling and Numerical Analysis*, 46(6):1447–1465, 2012.

[12] J. Solsvik and H. A. Jakobsen. Evaluation of weighted residual methods for the solution of a population balance model describing bubbly flows: the least-squares, Galerkin, tau, and orthogonal collocation methods. *Industrial and Engineering Chemistry Research*, 52(45):15988–16013, 2013.

[13] Y. P. Kim and J. H. Seinfeld. Simulation of multicomponent aerosol dynamics. *Journal of Colloid and Interface Science*, 149(2):425–449, 1991.

[14] S. Kumar and D. Ramkrishna. On the solution of population balance equations by discretization-I. A fixed pivot technique. *Chemical Engineering Science*, 51(8):1311–1332, 1996.

[15] S. Kumar and D. Ramkrishna. On the solution of population balance equations by discretization-II. A moving pivot technique. *Chemical Engineering Science*, 51(8):1333–1342, 1996.

[16] J. Kumar, M. Peglow, G. Warnecke, and S. Heinrich. An efficient numerical technique for solving population balance equation involving aggregation, breakage, growth, and nucleation. *Powder Technology*, 182(1):81–104, 2008.

[17] J. Kumar, M. Peglow, G. Warnecke, and S. Heinrich. The cell average technique for solving multi-dimensional aggregation population balance equations. *Computers and Chemical Engineering*, 32(8):1810–C1830, 2008.

[18] S. E. Pratsinis. Simultaneous nucleation, condensation, and coagulation in aerosol reactors. *Journal of Colloid and Interface Science*, 124(2):416–427, 1988.

[19] R. McGraw. Description of aerosol dynamics by the quadrature method of moments. *Aerosol Science and Technology*, 27(2):255–265, 1997.

[20] D. L. Marchisio and R. O. Fox. Solution of population balance equations using the direct quadrature method of moments. *Journal of Aerosol Science*, 36(1):43–73, 2005.

[21] M. Frenklach. Method of moments with interpolative closure. *Chemical Engineering Science*, 57(12):2229–2239, 2002.

[22] M. Z. Yu, J. Z. Lin, and T. L. Chan. A new moment method for solving the coagulation equation for particles in Brownian motion. *Aerosol Science and Technology*, 42(9):705–713, 2008.

[23] J. A. Shohat and J. D. Tamarkin. *The Problem of Moments*. American Mathematical Society, Providence, Rhode Island, revised edition, 1970.

[24] Douglas L. Wright Jr. Numerical advection of moments of the particle size distribution in Eulerian models. *Journal of Aerosol Science*, 38:352–369, 2007.

[25] D. T. Gillespie. An exact method for numerically simulating the stochastic coalescence process in a cloud. *Journal of the Atmospheric Sciences*, 32(10):1977–1989, 1975.

[26] K. Zhou, Z. He, M. Xiao, and Z. Zhang. Parallel Monte Carlo simulation of aerosol dynamics. *Advances in Mechanical Engineering*, 2014: 435936, 2014.

[27] K. Rajamani, W. T. Pate, and D. J. Kinneberg. Time-driven and event-driven Monte Carlo simulations of liquid-liquid dispersions: a comparison. *Industrial and Engineering Chemistry Fundamentals*, 25:746–752, 1986.

[28] N. Ahmed, G. Matthies, and L. Tobiska. Finite element methods of an operator splitting applied to population balance equations. *Journal of Computational and Applied Mathematics*, 236(6):1604–1621, 2011.

[29] N. Ahmed, G. Matthies, and L. Tobiska. Stabilized finite element discretization applied to an operator-splitting method of population balance equations. *Applied Numerical Mathematics*, 70(0):58–79, 2013.

[30] F. Anker, S. Ganesan, V. John, and E. Schmeyer. A comparative study of a direct discretization and an operator-splitting solver for population balance systems. *Computers and Chemical Engineering*, 75:95–104, 2015.

[31] K. Zhou, A. Attili, A. Alshaarawi, and F . Bisetti. Simulation of aerosol nucleation and growth in a turbulent mixing layer. *Physics of Fluids*, 26:065106, 2014.

[32] K. Zhou and Z. He. Monte Carlo simulation of aerosol evolution in a planar mixing layer. *International Journal of Numerical Methods for Heat & Fluid Flow*, 24(8):1769–1781, 2014.

[33] R. I. McLachlan and G. R. W. Quispel. Splitting methods. *Acta Numerica*, 11:341–434, 2002.

[34] V. I. Kalikmanov. *Nucleation Theory*. Springer, Dordrecht, The Netherlands, 2013.

[35] R. Becker and W. Döring. Kinetische Behandlung der Keimbildung in übersättigten Dämpfern. *Annalen der Physik (Leipzig)*, 24(8):719–752, 1935.

[36] B. D. Flury. Acceptance-reject ion sampling made easy. *SIAM Review*, 32(3):474–476, 1990.

[37] G. S. Fishman. *Monte Carlo: Concepts, Algorithms, and Applications*. Springer-Verlag, New York, 1996.

[38] H. Jia and K. Li. A third accurate operator splitting method. *Mathematical and Computer Modelling*, 53(1–2):387–396, 2011.

[39] Z. He, K. Zhou, M. Xiao, and F. Wei. Simulation of soot size distribution in a counterflow flame. *International Journal of Chemical Reactor Engineering*, 12(2):1–7, 2014.

Chapter 2

Methods of Moments for Resolving Aerosol Dynamics

Mingzhou Yu and Liu Yueyan

Additional information is available at the end of the chapter

http://dx.doi.org/10.5772/65565

Abstract

The study on aerosol dynamics processes, such as formation of nano/microscale aerosol particle and its subsequent growth in quiescent or evolving flows, has received much attention from both chemical engineering and atmospheric environment communities. The suitable theoretical method for resolving aerosol dynamical processes is widely known as population balance modeling (PBM), which is based on solving the population balance equation (PBE) in terms of particle number concentration. The study on the solution of the PBE has undergone rapid development in last several decades. In this chapter, the development of the method of moments for solving the PBE is presented. Three main methods of moments, including the Taylor series expansion method of moments, log-normal method of moments, and quadrature method of moments, are discussed.

Keywords: aerosol dynamics, method of moments, population balance equation

1. Introduction

Aerosol particles usually refer to fine particles in air whose size is smaller than micrometer [1]. This type of particles can be found in a wide range of industrial and natural phenomena such as nanoparticle synthesis [2, 3], aerosol sciences and air pollution [4–7], contamination control in the microelectronics and pharmaceuticals industries [8], and diesel particulate formation [9]. The dynamics characteristics of size of these particles spans from free molecular size regime much less than Kolmogorov length scale to inertial range. Due to Brownian motion, aerosol varies greatly in the degree of stability, even though the aerosol flow convection transport is not involved. Although this type of multiphase system widely emerges in industries and our surroundings, some key issues including the conversion from gas to particle and the subsequent particle growth affected by the surroundings remain unresolved [10]. Unlike some common techniques used in the multiphase flow community, the task of the study on aerosol dynamics is to grasp the interaction between the dispersed particles and the carrier phase and also to obtain the fundamentals of internal processes such as nucleation, condensation, coagulation, and breakage [11].

If the particle formation and subsequent growth are studied theoretically, the theory should cover from simple kinetics theory to continuum theory. The Stokes' law needs to be modified as applied in this field because aerosol particles are usually smaller than molecular mean free path [1]. Some common methodologies in multiphase flow community, such as Euler-Lagrange and Euler-Euler, are unsuitable to be used in this field. In fact, in the first decades of last century, studies on dynamics of micro- and nanoscale particles are always the focus of physical science [12]. With further requirements of modern industrial nanoparticle synthesis and atmospheric observation, the study on interaction between fine particles and the surrounding becomes more and more important, and thus, it needs to combine the methodology in modern multiphase flow theory with aerosol dynamics to resolve complicated micro- and nanoscale particle multiphase problems [13]. The object of these studies is to capture the property, behavior, and physical principle of aerosol particles in air and further apply this knowledge to their measurement and control.

Besides convection and diffusion transport, the evolution of aerosol particle dynamics arises mainly from internal mechanisms, including homogeneous or heterogeneous nucleation, condensation, coagulation, and breakage. Among these internal mechanisms, coagulation occurs most commonly, but it is yet the most difficult to be treated from the viewpoint of mathematics, because the correlation among all the particles must be concerned separately [14]. Since the pioneering work of Smoluchowski in 1917, the mean-field theory has been introduced in aerosol collision problems and has been basis for numerous theoretical applications. These applications include the derivation of coagulation kernel under different mechanisms, the solution of the governing equations within the Smoluchowski mean-field theory, and the application of Smoluchowski mean-field theoretical model to predict the behavior of aerosol multiphase system. In fact, within the Smoluchowski mean-field theory, some important phenomena such as self-preserving distribution [15–17], and gelation or asymptotic behavior [18–21] have been thoroughly studied using state-of-the-art technologies such as the method of moments (MOM) [17, 19–23], sectional method (SM) [24–30], stochastic particle method [31–35], and Monte Carlo method [36–42]. The method of moments is more widely used than other methods due to that it requires the least requirement for computational cost as well as the relative simplicity of implementation [36].

Although the method of moments has become a powerful tool for investigating aerosol chemistry physical processes since it was first used in aerosol community by Hulburt and Katz [22], the closure of governing equations in terms of kth moment is not easy to achieve. Up to now, there have been five main techniques proposed to achieve the closure of moment equations, namely the Taylor series expansion MOM (TEMOM) [43], the predefined size distributed method such as log-normal MOM (log MM) [44, 45] and Gamma MOM [46], Gaussian quadrature MOM (QMOM) and its variants [47, 48], pth-order polynomial MOM [49], and MOM with interpolative closure (MOMIC) [23]. In recent years, the PBM scheme, which couples the PBE with the Computational Fluid Dynamics (CFD), has been increasingly received attention, and accordingly, it is possible to simultaneously capture the details of the fluid flow and transport, the evolution of the particle size distribution and complex chemical kinetics. For any techniques, consuming computational cost has to be concerned as the solution of Navier-Stokes equations is involved. Thus, the efficiency of solving the PBE in the implementation of the PBM is another important issue besides the accuracy of numerical calculation. Although the QMOM and its variants are the most used scheme for solving the PBE today, it shows disadvantage in efficiency as compared to the TEMOM and log MM. It needs to note here in the log MM, and the log-normal size distribution has to be employed in the construction of the model, which inevitably weakens its reliability and capability for solving the PBE. It is necessary to construct a new approach with respect to moment equation, which is easy to implement with low computational cost like the log MM and has not the prior requirement for particle size spectrum like the QMOM, to adapt to the requirement of modern complicated particulate industries. In particular, there needs a suitable technique capable of providing explicit moment governing equations for further asymptotic analysis for the PBE [50]. In order to accomplish it, a new promising method of moments based on Taylor series expansion technique has been proposed and successfully applied to resolve some aerosol engineering problems [43]. Relative to the QMOM and log MM, the TEMOM has advantage to give explicit moment governing equations, making it suitable as basis equation for further analytical solution or asymptotic solution of the PBE.

This chapter is outlined as follows: In Section 2, the review of the PBE as well as its solution is presented, in which three main techniques applied for solving the PBE, namely the method of moments, sectional method, and Monte Carlo method, are briefly presented; the method of moments is highlighted in Section 3, where three predominated methods of moments, including the QMOM, log MM, and TEMOM, are presented separately.

2. Methods for solving the PBE

The study of PBE dates back to 1917 when famous polish scientist, Smoluchowski, first established the discrete governing equation for colloid coagulation, that is, Smoluchowski equation [12]; then, Smoluchowski equation was further developed by Müller in its integral-differential form, which finally becomes the basis equation of the PBM [51]. Today, the Smoluchowski equation has developed from its original version only accounting for coagulation to the present version accounting to almost all aerosol dynamics, including external processes such as particle convection and diffusion transport in air, and internal processes such as nucleation, coagulation, condensation, and breakage [52, 53]. Without loss of generality, the general form of a PBE, accounting for both external and internal processes, can be expressed as: where $n(v,x_i,t)$ is the particle number density for particle volume v, location x, and time t; u_j is the particle velocity; u_{th} is the thermophoresis velocity; D_B is the Brownian diffusion coefficient; G_r is the particle surface growth rate; J is the nucleation rate for the critical monomer volume v^*; β is coagulation kernel between two particles; and a and b are parameters accounting for the breakage rate associated with the turbulence shear force. Equation (1) encompasses almost all physicochemical processes of aerosol with a size smaller than approximately 1 μm and therefore is reliable for studying aerosol dynamics. In particular, an inherent advantage is that it can be coupled with the Navier-Stokes equation.

$$\begin{aligned}&\frac{\partial n(v,x_i,t)}{\partial t}+\frac{\partial\left(u_j n(v,x_j,t)\right)}{\partial x_j}+\frac{\partial\left((u_{th})_j n(v,x_j,t)\right)}{\partial x_j}\\&=\frac{\partial}{\partial x_j}\left(D_B\frac{\partial n(v,x_j,t)}{\partial x_j}\right)+\frac{\partial\left(G_r n(v,x_i,t)\right)}{\partial v}+J\left(v^*,x_i,t\right)\delta\left(v-v^*\right)\\&+\frac{1}{2}\int_{v^*}^{v}\beta\left(v-v',v'\right)n\left(v-v',t\right)n\left(v',x_i,t\right)\mathrm{d}v'\\&-n(v,t)\int_{v^*}^{\infty}\beta\left(v,v'\right)n\left(v',x_i,t\right)\mathrm{d}v'+\int_{v}^{\infty}a\left(v'\right)b\left(v|v'\right)n\left(v',t\right)\mathrm{d}v'\\&-a(v)n(v,t)+\ldots,\end{aligned}\tag{1}$$

It needs to note here in Eq. (1), the coordinate (i.e., particle volume v) of particle number concentration function might be other quantity, such as particle surface area or charge number, and coordinate number might be more, depending on the specific requirement of study. Even only the particle volume is selected as coordinate of particle number concentration function, the direct numerical solution of Eq. (1) is intractable for most applications due to the extreme large number of independent variables, and it should be further modified using suitable mathematical techniques. To solve the Eq. (1) numerically or analytically, several schemes by different researchers, including the method of moments, sectional method, and stochastic particle method, have been proposed and evaluated. Both advantages and disadvantages of these three methods have been compared in many review articles [36, 53]. In case the coagulation kernel is simplified with homogeneous assumption, the analytical solution of PBE can be achieved [54–57]. The analytical method has been used to study nanoparticle dynamics in an experimental chamber [58].

Because of relative simplicity of implementation and low computational cost, the method of moments has been extensively used to solve the PBE. In the application of this method, the fractal moment variables inevitably appear in the transfer from the PBE to moment governing equation, which needs to be further treated with different techniques. Due to the low requirement for computational cost, in the last decade, the combination of method of moments and Computational Fluid Dynamics (CFD) technique has been an emerging research field; the task is to investigate the temporal and spatial evolution of nanoparticles under turbulent conditions.

The information of particle size distribution is lost due to the integral in the transfer from the PBE to moment governing equations in the method of moments, and thus this method is unable to trace the evolution of particle size distribution (PSD) if the reconstruct technique of PSD is not implemented [59]. The sectional method, which divides the PBE into a set of size classes, overcomes the limit of method of moments in tracing the PSD. This method was usually used as an exact solution to validate the method of moments [60] and also widely applied in studies on the evolution of particle size distribution at engineering conditions due to the different dynamical processes including coagulation, condensation, gas-particle conversion, etc.

An alternative to sectional method and method of moments for solving the PBE is stochastic particle method (or Monte Carlo method) [31–35]. The application of this method, however, is limited because of low efficiency. This method has advantage to capture the evolution of particle size distribution physically and also can be used to obtain some key kernels for aerosol dynamics such as coagulation. Up to now, there have been lots of versions of Monte Carlo methods for solving the PBE, but the coupling between the Computational Fluid Dynamics (CFD) and Monte Carlo method is still limited.

3. Method of moments

The evolution of aerosol particle behavior arises from the interaction between particles and surrounding air; these small particles share energy with gas molecules and exhibit Brownian motion. With the exception of convection and diffusion transport, the nucleation and condensation mechanisms account for the mass or energy transfer from gas vapors to particle system, while coagulation and breakage mechanisms account for particle number variance. In theory, all the mechanisms can be defined as a function of time from a macroscopic probability view. Within the Smoluchowski mean-field theory, all aerosol dynamics processes can be invoked into the particle general dynamic equation, that is, the PBE. Except nucleation and condensation mechanisms, some difficulties are from closure of unresolved moment, which limits the application of moment methodology in this field. This is not only because there is nonintegral form in coagulation or breakage kernel, but also because there are much more specific mechanisms to concern.

Typically, turbulence is the driving force for particle radii of about 1–10 μm, while smaller particles are driven by Brownian motion and larger particles by differential sedimentation [61]. Here, coagulation receives much more attention by scientists than other dynamics processes such as nucleation and condensation because coagulation mechanism is harder to dispose in mathematics [14, 62]. The collision between interparticles is assumed to be instantaneous with spherical shape, while the mathematical description for agglomerate structures composed of noncoalescing spheres should be specially disposed [17, 63–65].

3.1. Closure of moment equations

The key task of the method of moments for solving the PBE is to convert the PBE to moment governing equations, during the conversion some approximations have to be employed to achieve the closure of the moment governing equations. Once the closed moment governing equations were established, they can be solved by some common numerical techniques, such as the fourth-order Runge-Kutta method. In the past, some techniques have been proposed to implement the method of moments. In this section, we only focus on coagulation and present how these different methods of moments are implemented for resolving this issue.

In order to represent the evolution of particle number, it is necessary to define particle concentration as a function of time and particle volume. This disposition was first proposed by Smoluchowski [12] for coagulation in dilute electrolytes, which has been basis for solving micro- and nanoparticle multiphase problems in modern aerosol or colloid science. The integral form of Smoluchowski equation is: where $n(v,t)\,dv$ is the number of particles whose volume is between v and $v+dv$ at time t, and $\beta(v,v')$ is the collision kernel for two particles of volumes v and v'. If the method of moments is used, the general disposition for this problem is to convert Eq. (2) into an ordinary differential equation with respect to the moment m_k. The conversion involves multiplying Eq. (2) by v^k and then integrating over the entire size distribution, and finally the converted moment equation based on the size distribution is obtained where

$$\frac{\partial n(v,t)}{\partial t}=\frac{1}{2}\int_0^v \beta(v-v',v')n(v-v',t)n(v',t)\mathrm{d}v'-n(v,t)\int_0^\infty \beta(v,v')n(v',t)\mathrm{d}v', \quad (2)$$

$$\frac{\mathrm{d}m_k}{\mathrm{d}t}=\frac{1}{2}\int_0^\infty\int_0^\infty \kappa(v,v',k)n(v,t)n(v',t)\mathrm{d}v\mathrm{d}v', \quad (3)$$

$$\kappa(v,v',k)=\left[(v+v')^k-v^k-v'^k\right]\beta(v,v').$$

During the implementing conversion from the Smoluchowski equation, or PBE, to the moment governing equation, a definition for m_k should be

$$m_k=\int_0^\infty v^k n(v)\mathrm{d}v. \quad (4)$$

The key task of all methods of moments, including the TEMOM, QMOM, MOMIC, and log MM, is to convert the integral term on the right hand of Eq. (3) to polynomials, and thereby, the numerical calculation can proceed. In this chapter, only the TEMOM, log MM, and QMOM will be presented. If readers have interests to other methods of moments, they are recommended to read articles for the MOMIC [23], the Gamma method of moments [66], and the DQMOM [48].

3.2. Taylor series expansion method of moments

The TEMOM was first proposed in 2008 in its numerical version for dealing with coagulation due to the Brownian motion [43]; since then, it was further developed [60, 64, 67]. On the basis of governing equations obtained from the TEMOM, researchers have found that it is easy to achieve analytical and asymptotic solutions of the PBE [19, 20, 56, 57, 68–70]. The TEMOM has been applied in many aerosol-related problems and has also successfully been used for the realistic environmental and engineering problems where multidynamics are involved [71, 72]. However, all of these quoted studies were only taken into consideration for three-order Taylor series expansion using integer moment sequence. The recent study shows that this kind of solution leads to shortcoming of the existing TEMOM, that is, the initial geometric standard deviation is limited, and the fractional moment at an initial stage cannot be accurately captured. Both shortcomings indeed greatly weaken the capability of the TEMOM. To overcome the shortcoming of the TEMOM in this aspect, a generalized TEMOM was currently proposed [60], in which the accuracy of numerical calculation is increased with increasing the orders of Taylor series expansion.

Here, we select coagulation in the continuum-slip regime as an example to present how the TEMOM is implemented. The coagulation kernel for agglomerates in the continuum-slip is [64] where v_{p0} is the volume of primary particles, $B_2 = 2k_bT/3\mu$, $\psi = \lambda A/(3/4\pi)^{1/3}$, $A = 1.591$, $f = 1/D_f$. To implement the TEMOM, we need to substitute Eq. (5) in Eq. (3) and then multiply v^k on both sides, and we can obtain the following expression, where

$$\beta\left(v,v'\right) = B_2\left(\frac{1}{v^f}+\frac{1}{v'^f}\right)\left(v^f+v'^f\right) + B_2\psi v_{p0}^{f-\frac{1}{3}}\left(\frac{1}{v^{2f}}+\frac{1}{v'^{2f}}\right)\left(v^f+v'^f\right), \tag{5}$$

$$\begin{cases} \dfrac{dm_0}{dt} = -\dfrac{B_2}{2}\displaystyle\int_0^\infty\int_0^\infty\left(\xi_1+\psi v_{p0}^{f-\frac{1}{3}}\xi_2\right)n\left(v,t\right)n\left(v',t\right)dvdv' \\ \dfrac{dm_1}{dt} = 0 \\ \dfrac{dm_2}{dt} = -\dfrac{B_2}{2}\displaystyle\int_0^\infty\int_0^\infty\left(\varsigma_1+\psi v_{p0}^{f-\frac{1}{3}}\varsigma_2\right)n\left(v,t\right)n\left(v',t\right)dvdv' \end{cases} \tag{6}$$

$$\xi_1 = 2 + v^f v'^{-f} + v^{-f} v'^f$$

$$\xi_2 = v^{-f} + v'^{-f} + v^{-2f} v'^f + v^f v'^{-2f}$$

$$\varsigma_1 = 4vv' + 2v^{1+f}v'^{1-f} + 2v^{1-f}v'^{1+f}$$

$$\varsigma_2 = 2v^{1+f}v'^{1-2f} + 2v^{1-2f}v'^{1+f} + 2v^{1-f}v' + 2vv'^{1-f}.$$

Once the definition for k -th moment, m_k, shown in Eq. (4), is introduced, we can obtain the following expression for moment governing equation, where

$$\begin{cases} \dfrac{dm_0}{dt} = -\dfrac{B_2}{2}(\xi_1^* + \psi v_{p0}^{f-\frac{1}{3}}\xi_2^*) \\ \dfrac{dm_1}{dt} = 0 \\ \dfrac{dm_2}{dt} = -\dfrac{B_2}{2}(\varsigma_1^* + \psi v_{p0}^{f-\frac{1}{3}}\varsigma_2^*) \end{cases} \quad (7)$$

$$\xi_1^* = 2m_0m_0 + 2m_f m_{-f}$$

$$\xi_2^* = 2m_{-f}m_0 + 2m_f m_{-2f}$$

$$\varsigma_1^* = 4m_1m_1 + 4m_{1+f}m_{1-f}$$

$$\varsigma_2^* = 4m_{1-f}m_1 + 4m_{1+f}m_{1-2f}.$$

It is obvious that Eq. (7) is not closed due to the appearance of some unexpected variables, such as m_{-f} and m_f. In the TEMOM, approximated functions are used to replace these unexpected variables, such as third-order Taylor series expansion function,

$$m_k = u_0^{k-2}\left(\frac{k^2-k}{2}\right)m_2 + u_0^{k-1}\left(-k^2+2k\right)m_1 + u_0^{k}\left(\frac{2+k^2-3k}{2}\right)m_0. \quad (8)$$

As the function shown in Eq. (8) is applied in Eq. (7), the final closed moment governing equation can be obtained where

$$\begin{cases} \dfrac{dm_0}{dt} = -\dfrac{B_2}{2}\left(\eta_1 + \psi v_{p0}{}^{f-\frac{1}{3}}\eta_2\right) \\ \dfrac{dm_1}{dt} = 0 \\ \dfrac{dm_2}{dt} = -\dfrac{B_2}{2}\left(\eta_3 + \psi v_{p0}{}^{f-\frac{1}{3}}\eta_4\right) \end{cases} \tag{9}$$

$$\eta_1 = \frac{-m_0^2(-5f^2m_1^4 + f^4m_1^4 + 8m_1^4 + f^4m_2^2m_0^2 - 2f^4m_2m_1^2m_0 + 6f^2m_2m_1^2m_0 - f^2m_2^2m_0^2)}{4m_1^4}$$

$$\eta_2 = \frac{-m_0{}^{2+f}(\psi_4 f^4 + \psi_3 f^3 + \psi_2 f^2 + \psi_1 f^1 + \psi_0)}{2m_1{}^{4+f}}$$

$$\eta_3 = \frac{-5f^2m_1^4 + f^4m_1^4 + 8m_1^4 + f^4m_2^2m_0^2 - 2f^4m_2m_1^2m_0 + 6f^2m_2m_1^2m_0 - f^2m_2^2m_0^2}{2m_1^2}$$

$$\eta_4 = \frac{m_0{}^f(\psi_4 f^4 - \psi_3 f^3 + \psi_2 f^2 - \psi_1 f^1 + \psi_0)}{m_1{}^{2+f}}$$

$$\psi_4 = 2m_2{}^2m_0{}^2 + 2m_1{}^4 - 4m_2m_0m_1{}^2$$

$$\psi_3 = -m_2{}^2m_0{}^2 - m_1{}^4 + 2m_2m_0m_1{}^2$$

$$\psi_2 = -m_2{}^2m_0{}^2 - 7m_1{}^4 + 8m_2m_0m_1{}^2$$

$$\psi_1 = -2m_1{}^4 + 2m_2m_0m_1{}^2$$

$$\psi_0 = 4m_1{}^4$$

Equation (9) is a system of first-order ordinary differential equations, all the right terms are denoted by the first three moments m_0, m_1, and m_2, and thus, this system can be automatically closed. It is clear in the derivation that no any physical assumption for the particle size distribution is introduced, making the TEMOM has more solid foundation in mathematics relative to the log MM.

The TEMOM was further developed to a much more general version, that is, generalized TEMOM. This newly developed version has some advantages as compared to old one. The new generalized TEMOM successfully overcomes the shortcomings of the old version whose geometric standard deviation must be less than a certain value. In addition, the accuracy of numerical calculation for capturing fractional moments at an initial stage can be largely increased. In the generalized TEMOM, the closure function shown in Eq. (8) is changed, in which the moment sequence is composed not only by integer moments but also by fractional moments. Thus, higher-order Taylor series expansion can be achieved for the closure function, making it much more accurate function.

3.3. Log-normal method of moments

Unlike the TEMOM, the closure of moment governing equations in the log MM is achieved by assuming the PSD to log-normal size distributions. This method was first investigated by Cohen and Vaughan whose work covers both Brownian and gravitational coagulation [73]. This work forms the basis of the computer code HAARM and aerosol dynamics model MAD [74], and the latter finally became key part in some atmospheric forecast models, such as WRF/chem. Thanks to works from scientists, including Lee [44, 75], and Pratsinis [14], the log MM becomes one of several main methods of moments today, and especially, it has been applied in computational fluid dynamics software, Fluent.

Similar to the TEMOM, the log MM also requires to first obtain the moment equations as shown in Eq. (7). Then, the task is to use its own closure function to achieve the closure of moment equations. In this method, the closure function is obtained on basis of the log-normal size distribution assumption, which has the following expression, where

$$m_k = m_0 {v_g}^k \exp\left(\frac{9}{2}k^2 \ln^2 \sigma\right) \tag{10}$$

$$v_g = \frac{{m_1}^2}{{m_0}^{3/2}{m_2}^{1/2}}$$

$$\ln^2 \sigma = \frac{1}{9}\ln\left(\frac{m_0 m_2}{{m_1}^2}\right).$$

In theory, once the PBE was converted to forms such as Eq. (7), in which unresolved moments are involved, it can be further numerically solved together with Eq. (10).

3.4. Quadrature method of moments

The QMOM and its variants such as DQMOM are regarded as the mostly used method of moments in the implementation of the PBM [47, 48]. This method achieves great success in that it has no any physical assumptions and has no requirement for the form of dynamics kernels. Thus, this method can deal with all relevant aerosol dynamics problems. Unlike the

http://dx.doi.org/10.5772/65565

log MM, this method does not need to first convert the PBE to unresolved moment equations, which needs further closed by approximated closure functions, such as Eqs. (8) and (10). Therefore, this method can be regarded as the most ideal scheme for solving the PBE if the numerical efficiency is not considered. In this method, the closure problem of the PBE is solved with a quadrature approximation. However, the weights and abscissas of the quadrature approximation need to be additionally obtained by suitable mathematical techniques, such as the product-difference algorithm. This increases the computational cost in contrast to the log MM and TEMOM. In fact, the numerical efficiency is similar important to the accuracy for simulation; especially, the coupling between the PBM and the computational fluid dynamics is considered.

Same as discussed in Sections 3.3 and 3.4, coagulation in the continuum-slip regime is selected as an example to present how the QMOM is implemented. To implement the QMOM, Eq. (3) needs to be disposed using Gaussian quadrature approximation as below, where v_i is the ith quadrature point, and ω_i is the corresponding weight in the quadrature formula. This method has insensitive form of kernel $\beta(v_i, v_j)$, and thus, extremely complicated kernel can be used. The method requires two times of quadrature point number to attain expected moments, for example, governing equations for moment m_0, m_1, …, m_5 need to be simultaneously solved as $N_Q = 3$.

$$\frac{dm_k}{dt} = \frac{1}{2}\sum_{i=1}^{N_Q}\sum_{i=1}^{N_Q}\left[\left(v_i + v_j\right)^k - v_i^k - v_j^k\right]\beta\left(v_i, v_j\right)\omega_i\,\omega_j \tag{11}$$

To implement the QMOM, the quadrature abscissas and weights are obtained from lower-order radial moment sequence by solving Eq. (11). During the implementation, the key task is to construct a symmetric tridiagonal matrix whose diagonal elements and off-diagonal elements are derived from the calculated moments. The symmetric tridiagonal matrix is diagonalized to obtain the abscissas and weights. It needs to note here that the number of abscissas and weights are dependent on N_Q. More details about the implementation of the QMOM are recommended to see the original work of this method.

Author details

Mingzhou Yu[1,2*] and Liu Yueyan[1]

*Address all correspondence to: yumingzhou1738@yahoo.com

1 China Jiliang University, Hangzhou, China

2 Key Laboratory of Aerosol Chemistry and Physics, Chinese Academy of Science, Xi'an, China

References

[1] S. K. Friedlander, Smoke, dust and haze: fundamentals of aerosol behavior, 2nd ed. John Wiley & Sons, Inc., New Jersey, 2000.

[2] B. Buesser and S. E. Pratsinis, Design of nanomaterial synthesis by aerosol processes, *Annu. Rev. Chem. Biomol. Eng.*, vol. 3, pp. 103–127, 2012.

[3] M. Yu, J. Lin, and T. Chan, Effect of precursor loading on non-spherical TiO_2 nanoparticle synthesis in a diffusion flame reactor, *Chem. Eng. Sci.*, vol. 63, no. 9, pp. 2317–2329, 2008.

[4] M. Kulmala, Atmospheric science. How particles nucleate and grow, *Science*, vol. 302, no. 5647, pp. 1000–1001, 2003.

[5] K. Savolainen, L. Pylkkänen, H. Norppa, G. Falck, H. Lindberg, T. Tuomi, M. Vippola, H. Alenius, K. Hämeri, J. Koivisto, D. Brouwer, D. Mark, D. Bard, M. Berges, E. Jankowska, M. Posniak, P. Farmer, R. Singh, F. Krombach, P. Bihari, G. Kasper, and M. Seipenbusch, Nanotechnologies, engineered nanomaterials and occupational health and safety – a review, *Saf. Sci.*, vol. 48, no. 8, pp. 957–963, 2010.

[6] A. Joonas Koivisto, M. Yu, K. Hämeri, and M. Seipenbusch, Size resolved particle emission rates from an evolving indoor aerosol system, *J. Aerosol Sci.*, vol. 47, pp. 58–69, 2012.

[7] M. Yu, A. J. Koivisto, K. Hämeri, and M. Seipenbusch, Size dependence of the ratio of aerosol coagulation to deposition rates for indoor aerosols, *Aerosol Sci. Technol.*, vol. 47, no. 4, pp. 427–434, 2013.

[8] S. K. Friedlander and D. Y. Pui, Emerging issues in nanoparticle aerosol science and technology experimental methods and instrumentation, *J. Nanoparticle Res.*, no. 4, pp. 313–320, 2004.

[9] P. Kumar, B. R. Gurjar, A. S. Nagpure, and R. M. Harrison, Preliminary estimates of nanoparticle number emissions from road vehicles in megacity Delhi and associated health impacts, *Environ. Sci. Technol.*, vol. 45, no. 13, pp. 5514–5521, 2011.

[10] J. Kirkby, J. Duplissy, K. Sengupta, C. Frege, H. Gordon, C. Williamson, M. Heinritzi, M. Simon, C. Yan, J. Almeida, J. Tröstl, T. Nieminen, I. K. Ortega, R. Wagner, A. Adamov, A. Amorim, A. K. Bernhammer, F. Bianchi, M. Breitenlechner, S. Brilke, X. Chen, J. Craven, A. Dias, S. Ehrhart, R. C. Flagan, A. Franchin, C. Fuchs, R. Guida, J. Hakala, C. R. Hoyle, T. Jokinen, H. Junninen, J. Kangasluoma, J. Kim, M. Krapf, A. Kürten, A. Laaksonen, K. Lehtipalo, V. Makhmutov, S. Mathot, U. Molteni, A. Onnela, O. Peräkylä, F. Piel, T. Petäjä, A. P. Praplan, K. Pringle, A. Rap, N. A. D. Richards, I. Riipinen, M. P. Rissanen, L. Rondo, N. Sarnela, S. Schobesberger, C. E. Scott, J. H. Seinfeld, M. Sipilä, G. Steiner, Y. Stozhkov, F. Stratmann, A. Tomé, A. Virtanen, A. L. Vogel, A. C. Wagner, P. E. Wagner, E. Weingartner, D. Wimmer, P. M. Winkler, P. Ye, X. Zhang, A. Hansel, J.

Dommen, N. M. Donahue, D. R. Worsnop, U. Baltensperger, M. Kulmala, K. S. Carslaw, and J. Curtius, Ion-induced nucleation of pure biogenic particles, *Nature*, vol. 533, no. 7604, pp. 521–526, 2016.

[11] M. Yu and J. Lin, Nanoparticle-laden flows via moment method: a review, *Int. J. Multiph. Flow*, vol. 36, no. 2, pp. 144–151, 2010.

[12] M. von Smoluchowski, Versuch einer mathematischen Theorie der Koagulationskinetik kolloider Lösungen, *Z. Phys. Chem.*, vol. 92, no. 2, pp. 129–168, 1917.

[13] D. Ramkrishna and M. R. Singh, Population balance modeling: current status and future prospects, *Annu. Rev. Chem. Biomol. Eng.*, vol. 5, pp. 123–146, 2014.

[14] S. Pratsinis, Simultaneous nucleation, condensation, and coagulation in aerosol reactors, *J. Colloid Interface Sci.*, vol. 124, no. 2, pp. 416–427, 1988.

[15] F. O. Ernst and S. E. Pratsinis, Self-preservation and gelation during turbulence-induced coagulation, *J. Aerosol Sci.*, vol. 37, no. 2, pp. 123–142, 2006.

[16] S. Vemury and S. E. Pratsinis, Self-preserving size distributions of agglomerates, *J. Aerosol Sci.*, vol. 26, no. 2, pp. 175–185, 1995.

[17] E. Goudeli, M. L. Eggersdorfer, and S. E. Pratsinis, Coagulation–agglomeration of fractal-like particles: structure and self-preserving size distribution, *Langmuir*, vol. 31, no. 4, pp. 1320–1327, 2015.

[18] Z. Varga, G. Wang, and J. Swan, The hydrodynamics of colloidal gelation, *Soft Matter*, vol. 11, pp. 9009–9019, 2015.

[19] M. L. Xie, Asymptotic solution of moment approximation of the particle population balance equation for Brownian agglomeration, *Aerosol Sci. Technol.*, vol. 49, no. 2, pp. 109–114, 2015.

[20] Z. Chen, J. Lin, and M. Yu, Asymptotic behavior of the Taylor-expansion method of moments for solving a coagulation equation for Brownian particles, *Particuology*, vol. 14, pp. 124–129, 2014.

[21] S. Park and K. Lee, Asymptotic particle size distributions attained during coagulation processes, *J. Colloid Interface Sci.*, vol. 233, no. 1, pp. 117–123, 2001.

[22] H. M. Hulburt and S. Katz, Some problems in particle technology: a statistical mechanical formulation, *Chem. Eng. Sci.*, vol. 19, no. 8, pp. 555–574, 1964.

[23] M. Frenklach, Method of moments with interpolative closure, *Chem. Eng. Sci.*, vol. 57, no. 12, pp. 2229–2239, 2002.

[24] H. Liu, M. Yu, Z. Yin, Y. Jiang, and M. Chen, Study on the evolution of nanoparticle size distribution due to continuous injection using the sectional method, *Int. J. Numer. Methods Heat Fluid Flow*, vol. 24, no. 8, pp. 1803–1812, 2014.

[25] M. M. Attarakih, C. Drumm, and H. J. Bart, Solution of the population balance equation using the sectional quadrature method of moments (SQMOM), *Chem. Eng. Sci.*, vol. 64, no. 4, pp. 742–752, 2009.

[26] M. C. Bruns and O. A. Ezekoye, Development of a hybrid sectional quadrature-based moment method for solving population balance equations, *J. Aerosol Sci.*, vol. 54, pp. 88–102, 2012.

[27] S. H. Park, A moving sectional model for the formation and growth of soot particles, pp. 1–13, 2003.

[28] J. D. Landgrebe and S. E. Pratsinis, A discrete-sectional model for particulate production by gas-phase chemical reaction and aerosol coagulation in the free-molecular regime, *J. Colloid Interface Sci.*, vol. 139, no. 1, pp. 63–86, 1990.

[29] M. Kostoglou, Extended cell average technique for the solution of coagulation equation, *J. Colloid Interface Sci.*, vol. 306, no. 1, pp. 72–81, 2007.

[30] F. Gelbard, Y. Tambour, and J. H. Seinfeld, Sectional representations for simulating aerosol dynamics, *J. Colloid Interface Sci.*, vol. 76, no. 2, pp. 541–556, 1980.

[31] K. Sabelfeld, Stochastic models for coagulation of aerosol particles in intermittent turbulent flows, *Math. Comput. Simul.*, vol. 47, nos. 2–5, pp. 85–101, 1998.

[32] F. E. Kruis, J. Wei, T. van der Zwaag, and S. Haep, Computational fluid dynamics based stochastic aerosol modeling: combination of a cell-based weighted random walk method and a constant-number Monte-Carlo method for aerosol dynamics, *Chem. Eng. Sci.*, vol. 70, pp. 109–120, 2012.

[33] E. Debry, B. Sportisse, and B. Jourdain, A stochastic approach for the numerical simulation of the general dynamics equation for aerosols, *J. Comput. Phys.*, vol. 184, no. 2, pp. 649–669, 2003.

[34] W. J. Menz, J. Akroyd, and M. Kraft, Stochastic solution of population balance equations for reactor networks, *J. Comput. Phys.*, vol. 256, pp. 615–629, 2014.

[35] S. L. Rani, R. Dhariwal, and D. L. Koch, A stochastic model for the relative motion of high Stokes number particles in isotropic turbulence, *J. Fluid Mech.*, vol. 756, pp. 870–902, 2014.

[36] M. Kraft, Modelling of particulate processes, *Kona*, vol. 23, no. 23, pp. 18–35, 2005.

[37] N. Morgan, C. Wells, M. Goodson, M. Kraft, and W. Wagner, A new numerical approach for the simulation of the growth of inorganic nanoparticles, *J. Comput. Phys.*, vol. 211, no. 2, pp. 638–658, 2006.

[38] J. Wei and F. E. Kruis, GPU-accelerated Monte Carlo simulation of particle coagulation based on the inverse method, *J. Comput. Phys.*, vol. 249, pp. 67–79, 2013.

[39] S. Khalili, Y. Lin, A. Armaou, and T. Matsoukas, Constant number Monte Carlo simulation of population balances with multiple growth mechanisms, *AIChE J.*, vol. 56, no. 12, pp. 3137–3145, 2010.

[40] H. Zhao and C. Zheng, A population balance-Monte Carlo method for particle coagulation in spatially inhomogeneous systems, *Comput. Fluids*, vol. 71, pp. 196–207, 2013.

[41] A. Vikhansky and M. Kraft, A Monte Carlo methods for identification and sensitivity analysis of coagulation processes, *J. Comput. Phys.*, vol. 200, no. 1, pp. 50–59, 2004.

[42] X. Hao, H. Zhao, Z. Xu, and C. Zheng, Population balance-Monte Carlo simulation for gas-to-particle synthesis of nanoparticles, *Aerosol Sci. Technol.*, vol. 47, no. 10, pp. 1125–1133, 2013.

[43] M. Yu, J. Lin, and T. Chan, A new moment method for solving the coagulation equation for particles in Brownian motion, *Aerosol Sci. Technol.*, vol. 42, no. 9, pp. 705–713, 2008.

[44] K. W. Lee, H. Chen, and J. A. Gieseke, Log-normally preserving size distribution for Brownian coagulation in the free-molecule regime, *Aerosol Sci. Technol.*, vol. 3, no. 1, pp. 53–62, 1984.

[45] S. E. Pratsinis, Simultaneous nucleation, condensation, and coagulation in aerosol reactors, *J. Colloid Interface Sci.*, vol. 124, no. 2, pp. 416–427, 1988.

[46] M. M. Williams, On the modified gamma distribution for representing the size spectra of coagulating aerosol particles, *J. Colloid Interface Sci.*, vol. 103, no. 2, pp. 516–527, 1985.

[47] R. McGraw, Description of aerosol dynamics by the quadrature method of moments, *Aerosol Sci. Technol.*, vol. 27, no. 2, pp. 255–265, 1997.

[48] D. L. Marchisio and R. O. Fox, Solution of population balance equations using the direct quadrature method of moments, *J. Aerosol Sci.*, vol. 36, no. 1, pp. 43–73, 2005.

[49] J. C. Barrett and J. S. Jheeta, Improving the accuracy of the moments method for solving the aerosol general dynamic equation, *J. Aerosol Sci.*, vol. 27, no. 8, pp. 1135–1142, 1996.

[50] M. L. Xie, Asymptotic behavior of TEMOM model for particle population balance equation over the entire particle size regimes, *J. Aerosol Sci.*, vol. 67, pp. 157–165, 2014.

[51] H. Müller, Zur allgemeinen Theorie ser raschen Koagulation, *Fortschrittsberichte über Kolloide und Polym.*, vol. 27, no. 6, pp. 223–250, 1928.

[52] V. Raman and R. O. Fox, Modeling of fine-particle formation in turbulent flames, *Annu. Rev. Fluid Mech.*, vol. 48, no. 1, pp. 159–190, 2016.

[53] U. Vogel, K. Savolainen, Q. Wu, M. van Tongeren, D. Brouwer, and M. Berges, *Handbook of Nanosafety: Measurement, Exposure and Toxicology*. Elsevier, Amsterdam, 2014.

[54] S. Park, R. Xiang, and K. Lee, Brownian coagulation of fractal agglomerates: analytical solution using the log-normal size distribution assumption, *J. Colloid Interface Sci.*, vol. 231, no. 1, pp. 129–135, 2000.

[55] E. Otto, H. Fissan, S. Park, and K. Lee, Log-normal size distribution theory of Brownian aerosol coagulation for the entire particle size range: part II—analytical solution using Dahneke's coagulation kernel, *J. Aerosol Sci.*, vol. 30, no. 1, pp. 17–34, 1999.

[56] M. Yu, J. Lin, J. Cao, and M. Seipenbusch, An analytical solution for the population balance equation using a moment method, *Particuology*, vol. 18, pp. 194–200, 2015.

[57] M. Yu, X. Zhang, G. Jin, J. Lin, and M. Seipenbusch, A new analytical solution for solving the population balance equation in the continuum-slip regime, *J. Aerosol Sci.*, vol. 80, pp. 1–10, 2015.

[58] M. Seipenbusch, A. Binder, and G. Kasper, Temporal evolution of nanoparticle aerosols in workplace exposure, *Ann. Occup. Hyg.*, vol. 52, no. 8, pp. 707–16, 2008.

[59] L. G. M. de Souza, G. Janiga, V. John, and D. Thévenin, Reconstruction of a distribution from a finite number of moments with an adaptive spline-based algorithm, *Chem. Eng. Sci.*, vol. 65, no. 9, pp. 2741–2750, 2010.

[60] M. Yu, Y. Liu, J. Lin, and M. Seipenbusch, Generalized TEMOM scheme for solving the population balance equation, *Aerosol Sci. Technol.*, vol. 49, no. 11, pp. 1021–1036, 2015.

[61] J. Chun and D. L. Koch, Coagulation of monodisperse aerosol particles by isotropic turbulence, *Phys. Fluids*, vol. 17, no. 2, pp. 027–102, 2005.

[62] R. R. Upadhyay and O. A. Ezekoye, Treatment of size-dependent aerosol transport processes using quadrature based moment methods, *J. Aerosol Sci.*, vol. 37, no. 7, pp. 799–819, 2006.

[63] M. M. Maricq, Coagulation dynamics of fractal-like soot aggregates, *J. Aerosol Sci.*, vol. 38, no. 2, pp. 141–156, 2007.

[64] M. Yu and J. Lin, Taylor-expansion moment method for agglomerate coagulation due to Brownian motion in the entire size regime, *J. Aerosol Sci.*, vol. 40, no. 6, pp. 549–562, 2009.

[65] M. Yu and J. Lin, Solution of the agglomerate Brownian coagulation using Taylor-expansion moment method, *J. Colloid Interface Sci.*, vol. 336, no. 1, pp. 142–149, 2009.

[66] M. M. R. Williams, Some topics in nuclear aerosol dynamics, *Prog. Nucl. Energy*, vol. 17, no. 1, pp. 01–52, 1986.

[67] Z. Chen, J. Lin, and M. Yu, A direct expansion method of moments for Brownian coagulation, *Phys. Scr.*, vol. 89, no. 12, pp. 125204, 2014.

[68] M. Yu, Y. Liu, G. Jin, and H. Jin, A new analytical solution for agglomerate growth undergoing Brownian coagulation, *Appl. Math. Model.*, vol. 40, pp. 5497–5509, 2016.

[69] M. Xie and Q. He, Analytical solution of TEMOM model for particle population balance equation due to Brownian coagulation, *J. Aerosol Sci.*, vol. 66, pp. 24–30, 2013.

[70] M. L. Xie, Asymptotic behavior of TEMOM model for particle population balance equation over the entire particle size regimes, *J. Aerosol Sci.*, vol. 67, pp. 157–165, 2014.

[71] M. Yu and J. Lin, Binary homogeneous nucleation and growth of water–sulfuric acid nanoparticles using a TEMOM model, *Int. J. Heat Mass Transf.*, vol. 53, no. 4, pp. 635–644, 2010.

[72] M. Yu, J. Lin, and T. Chan, Numerical simulation for nucleated vehicle exhaust particulate matters via the TEMOM/LES method, *Int. J. Mod. Phys. C*, vol. 20, no. 3, pp. 399–421, 2009.

[73] E. R. Cohen and E. U. Vaughan, Approximate solution of the equations for aerosol agglomeration, *J. Colloid Interface Sci.*, vol. 35, no. 4, pp. 612–623, 1971.

[74] E. R. Whitby and P. H. McMurry, Modal aerosol dynamics modeling, *Aerosol Sci. Technol.*, vol. 27, no. 6, pp. 673–688, 1997.

[75] K. Lee, Y. Lee, and D. Han, The log-normal size distribution theory for Brownian coagulation in the low Knudsen number regime, *J. Colloid Interface Sci.*, vol. 492, no. 188, pp. 486–492, 1997.

Chapter 3

Computational Fluid-Particle Dynamics Modeling for Unconventional Inhaled Aerosols in Human Respiratory Systems

Yu Feng, Zelin Xu and Ahmadreza Haghnegahdar

Additional information is available at the end of the chapter

http://dx.doi.org/10.5772/65361

Abstract

The awareness is growing of health hazards and pharmaceutical benefits of micro-/nano-aerosol particles which are mostly nonspherical and hygroscopic, and categorized as "unconventional" vs. solid spheres. Accurate and realistic numerical models will significantly contribute to answering public health questions. In this chapter, fundamentals and future trends of computational fluid-particle dynamics (CFPD) models for lung aerosol dynamics are discussed, emphasizing the underlying physics to simulate unconventional inhaled aerosols such as fibers, droplets, and vapors. Standard simulation procedures are presented, including reconstruction of the human respiratory system, CFPD model formulation, finite-volume mesh generation, etc. Case studies for fiber and droplet transport and deposition in lung are also provided. Furthermore, challenges and future directions are discussed to develop next-generation models. The ultimate goal is to establish a roadmap to link different numerical models, and to build the framework of a new multiscale numerical model, which will extend exposure and lung deposition predictions to health endpoints, e.g., tissue and delivered doses, by calculating absorption and translocation into alveolar regions and systemic regions using discrete element method (DEM), lattice Boltzmann method (LBM), and/or physiologically based pharmacokinetic (PBPK) models. It will enable simulations of extremely complex airflow-vapor-particle-structure dynamics in the entire human respiratory system at detailed levels.

Keywords: computational fluid-particle dynamics (CFPD) model, lung aerosol dynamics, unconventional inhaled aerosol, fiber, droplet, vapor, whole lung model, lattice Boltzmann method (LBM), compartmental modeling, physiologically based pharmacokinetic (PBPK) modeling

1. Introduction

Inhalable aerosols consist of multiple nano-to-micro-scale solid or liquid particulate matters (PMs) with dissolved or embedded compounds, as well as associated vapors. Instead of being spheres with constant diameters, most of these airborne PMs are nonspherical in shape, with everchanging sizes due to evaporation/condensation. Therefore, they are defined as "unconventional aerosol particles". Examples include fiber-like carbon nanotubes (CNTs) [1], hygroscopic cigarette particles [2–4], and pulmonary drug powders [5].

Both active and passive human exposures to ultrafine airborne aerosols are growing. Contributing factors include the increase in nanomaterial generation during product/device manufacturing, handling and use [6], consumptions of conventional and new tobacco products [2–4], as well as nanoparticle pulmonary drug development [7, 8]. Of great concern are the toxic health risks and therapeutic benefits induced by the inhalation. This is mainly because the small sizes facilitate ultrafine PM transport into deeper lung airways, translocation to the bloodstream crossing body membrane barriers, thus affecting organs and tissues in both pulmonary routes and systemic regions. For example, the transmission of respiratory infectious disease begins from the deposition of pathogens carried by airborne PM into the respiratory tract [9]. Since infection initiation is usually region specific, local dosages (i.e., lung deposited, tissue, and delivered doses) in pulmonary routes and systemic regions are essential for precise risk assessments. Another example is the worker exposure to airborne nanomaterials in the workplace [6]. Furthermore, pulmonary targeted drug delivery is favorable for both lung and systemic disease treatments, owing to the strong capability of the lung to absorb pharmaceuticals. Using ultrafine unconventional particles as drug carriers has also been promoted in medical device design, encapsulation, and drug delivery [7, 8, 10], because of their extraordinary capabilities to penetrate into deeper lung airways by virtue of the nonspherical shape or the hygroscopic growth characteristics. In this case, accurate local dosage information is also essential for the evaluation of drug delivery efficacy to targeted sites.

In light of the above discussion, the primary goal of relevant researches is to provide high-resolution data and enhance the fundamental understanding of the physical and chemical mechanisms for the dynamics of unconventional inhaled aerosols in human respiratory systems and systemic regions. Data of interests include: transport, deposition, dissolution, absorption, distribution, and clearance of unconventional aerosols in lung airways and systemic regions under various exposure conditions. These data will also help to answer the key question: "*What type of inhaled aerosols deposits where at what surface concentrations in subject-specific human respiratory systems under distinct breathing conditions, and what delivered dose into the site of interest in systemic regions?*"

Experiments or clinical tests are not able to provide such high-resolution data on human subjects, because of ethical reasons and lack of reliable measurements. Hence, developing and applying an accurate and realistic computer simulation model is desired. Different types of numerical models have been developed specifically to determine the fate of inhaled aerosols in human bodies, as has been reviewed recently by Phalen and Raabe [11]. To simulate particle trajectories, interactions, and lung deposition of unconventional inhaled aerosols, computational fluid-particle dynamics (CFPD) models based on Euler-Lagrange and Euler-Euler methods are the best options compared to other numerical models such as empirical and semiempirical ones [11]. Specifically, accurate prediction of the fate of such aerosols requires the model to be able to capture the effects of particle shape and size change on their trajectories before deposition. Based on more underlying physics and less extrapolations and simplifications, CFPD models are able to describe the fundamental behaviors of fluids and particles with high-resolution local data by solving conservation laws with constitutive equations [3, 4, 10]. Hence, these validated noninvasive, cost-effective, and accurate CFPD models will complement in vitro and in vivo studies on unconventional lung aerosol dynamics by simulating fluid-particle flows that are difficult to be reproduced experimentally. They will be able to deliver new perspectives of scientific knowledge and healthcare data for medical device improvements for effective and targeted drug deliveries, novel lung therapeutics, noninvasive disease diagnostic methodologies, and exposure health risk evaluations. Accordingly, CFPD models are expected to contribute to the following aspects:

1. Understand fundamental, physical, and chemical mechanisms that influence transport phenomena of different inhaled aerosol.
2. Virtualize corrective surgery leading to improved breathing patterns in abnormal airways.
3. Enhance evaluating capabilities of toxicologists on adverse health effects of inhaled particulate matters.
4. Design and engineer novel carriers for pulmonary drug delivery.
5. Establish guidelines for limiting workplace exposure to new nanoparticles and micron fibers, considering both short-term and long-term exposures.

This chapter is intended to serve as an overview of advanced CFPD models for unconventional inhaled aerosols and a demonstration of how to build and apply CFPD models on lung aerosol dynamics simulations. It also shows a roadmap to the next-generation multiscale lung dosimetry models. Specifically, fundamental mathematics of advanced CFPD modeling techniques is introduced in Section 2, focusing on predicting unconventional aerosol transport and deposition in human upper airways, i.e., nonspherical fiber, hygroscopic droplet, vapor mixture, etc. Basic steps of CFPD modeling are also discussed in this section. Section 3 presents applications of different advanced CFPD models with parametric sensitivity analyses on particle transport and deposition characteristics in airways. Challenges and future directions are discussed in Section 4 on developing next-generation multiscale numerical models, which will be able to simulate unconventional aerosol transport, deposition, and translocation from the complete pulmonary routes into systemic regions. Structured literature reviews on the above topics are also integrated in the following sections. Other reviews can be found in [12, 13].

2. CFPD models for unconventional lung aerosols transport

Computational fluid-particle dynamics (CFPD) incorporates computational fluid dynamics (CFD) theories with multiphase flow modeling techniques for particle dynamics in fluid flows, i.e., Euler-Euler and Euler-Lagrange methods [12, 13]. As shown in **Figure 1**, basic steps to proceed simulations of transport and deposition of inhaled aerosol in human respiratory systems using CFPD models include [11–13]: (1) reconstruction of the human respiratory system, (2) mesh generation and independence test, (3) formulation of governing equation system, (4) numerical discretization and solution, (5) model validation and revision, and (6) aerosol transport and deposition visualization and analyses with insights.

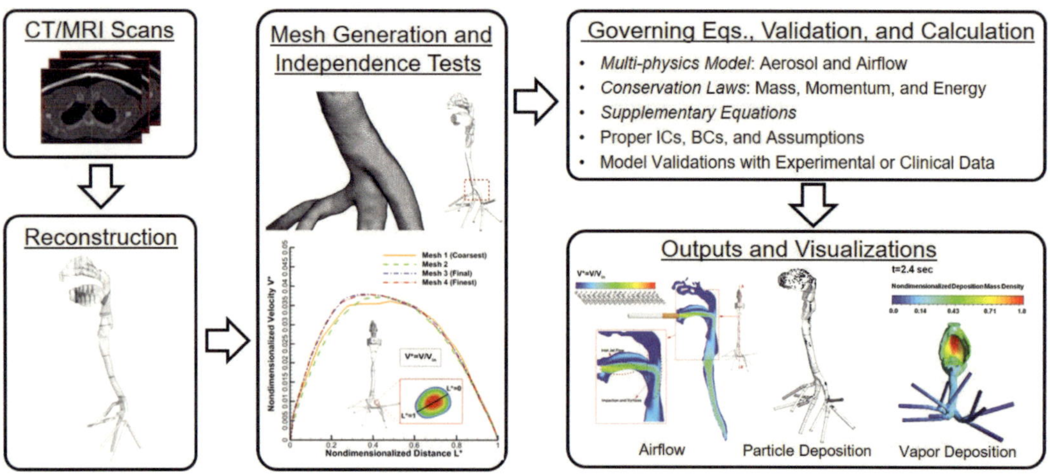

Figure 1. Basic steps to perform CFPD simulations for transport and deposition of aerosols in human respiratory systems.

The underlying assumptions of conventional CFPD models are that the aerosols are spherical, noninteracting, and constant in particulate diameter [12]. These assumptions induce inaccuracies in local deposition predictions when modeling the dynamics of unconventional aerosol particles, e.g., nonspherical particles and hygroscopic droplets. To accurately capture the transport and deposition of particulate matters, more physical and chemical principles must be considered, such as the rotational motion of nonspherical fibers, particle-particle collisions and coagulations, hygroscopic behaviors of droplets, as well as droplet-vapor phase transition, which strongly impact their trajectories and subsequent deposition patterns in human respiratory systems. Thus, advanced CFPD models have been developed and applied to modeling unconventional particle dynamics [2, 3, 10, 13]. In this section, procedures for CFPD modeling are discussed, followed by introducing governing equation systems of two advanced CFPD models for ellipsoidal fiber dynamics and multicomponent droplet-vapor aerosol transport in human respiratory systems.

2.1. Reconstruction of representative human respiratory systems

Since geometric variations have strong impacts on the airflow patterns leading to different particle trajectories, the geometric accuracy of human respiratory system is of great importance. Therefore, as the flow domain for CFPD modeling, the human respiratory system geometry must be precise in anatomy and physiology.

2.1.1. *Anatomy of human respiratory systems*

Human respiratory system consists of respiratory passages, lung lobes, and respiratory muscles (see **Figure 2**). Respiration includes three separate but related functions: ventilation (breathing), gas exchange (between air and blood in the lungs and between the blood and tissues), and oxygen utilization (by the tissues in the energy-liberating reactions of cell respiration). During the respiration process, airborne particulate matters are also inhaled and may settle in different locations. The internal walls of the human respiratory system are covered by mucus layers. As a filtering mechanism, mucus layers serve to trap and clear the PM carried by the lung airflow.

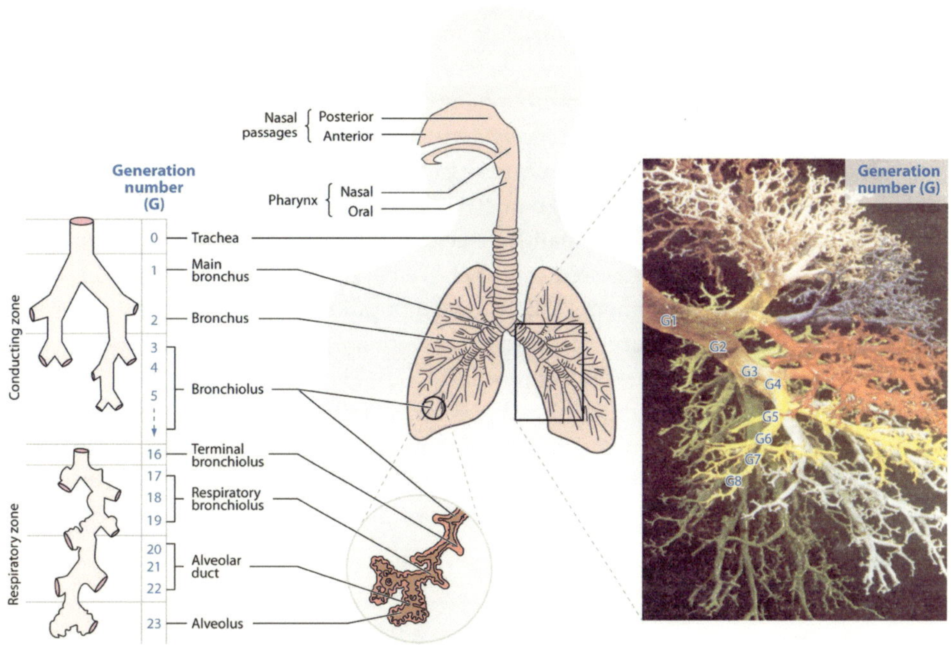

Figure 2. Schematics of human respiratory system (reprinted from [12]).

The human respiratory system ventilation path contains mouth, nose, pharynx, glottis, larynx, trachea, bronchi, bronchioles (including terminal bronchioles (generation 16 (G16)) and respiratory bronchioles (generation 17–19 (G17–G19)), and alveoli [14]. The path can be divided into two functional zones. The conducting zone includes all of the anatomical structures through which air passes before reaching the respiratory zone, i.e., mouth to terminal bronchioles (see **Figure 2**). The respiratory zone is the region where gas exchange occurs, which includes respiratory bronchioles and alveoli.

Human respiratory system can also be divided into upper airway and lower airway. The upper airway is all structures above the glottis; the lower airway is from the vocal cords to the lung. The upper airway include oral cavity, pharynx, larynx, and upper part of the trachea. Specifically, oral cavity can be considered as having the shape of a prolate spheroid that features different radius of curvatures along the sagittal and coronal planes. The shape of the oropharynx is more circular before the glottal region. In the lower airway, the lung consists of 23 generations, encompassing 2^{23} airways plus millions of alveoli of different shapes. There are about 300 million alveoli (0.25–0.5 mm in diameter) in the lung which provide a high surface area, i.e., 60–80 m^2 for diffusion of gases. Each alveolus is only one cell-layer thick, so that the total "air-blood barrier" is only two cells across (an alveolar cell and a capillary endothelial cell), which is approximately 2 μm.

Another way is to divide the human respiratory system into three parts [12]: the extrathoracic region (i.e., nose, mouth, and throat), the tracheobronchial part (i.e., trachea and bronchial tree), and the alveolar region (i.e., alveolar ducts and sacs).

2.1.2. Idealized human upper airway geometries

Idealized geometries are built up of simple geometric shapes but possess all the basic anatomical features of real human respiratory systems. At the early age, due to the limitations of computational power and imaging techniques handling subject-specific geometries, the use of idealized representative upper airway geometries is indispensable to investigate airflow and aerosol deposition patterns [15–19]. Idealized tracheobronchial geometries are usually created following airway dimensions measured by either Weibel [20] or Horsfield et al. [21]. They also follow different sets of measurements [22, 23]. Two popular idealized geometries are shown in **Figure 3(a)** and **(b)** [3, 18]. **Figure 3(a)** contains oral cavity, pharynx, larynx, and a symmetric, planar triple-bifurcation lung airway representing G0–G3 after Cheng et al. [22] and Weibel [20]. **Figure 3(b)** presents another idealized geometry with oral airways, nasal airways, and asymmetric triple bifurcation unit (TBU) representing G0 to G3 following Alberta-Finlay model [22] and Horsfield et al. [21]. Newly developed idealized human upper airway geometries emphasize more on matching average aerosol deposition data measured in different airway replicas [24].

Although idealized geometries are still beneficial for extensive fundamental transport mechanism studies and easier model validations with in-vitro measurements [2, 3, 10, 25], subject-specific respiratory system geometries are necessary to be employed for individualized risk assessment or disease treatment. The importance of airway geometry on micron-particle deposition was investigated and confirmed [26].

2.1.3. Subject-specific human upper airway geometries

Currently, more CAD-like subject-specific human upper airway geometry files are generated via DICOM file conversion using either specialized software or in-house codes. An example is shown in **Figure 4**. The geometry is from a 20-mm opening mouth inlet, to the tracheobronchial airways up to G9, which has been generated based on CT images. The subject for this

http://dx.doi.org/10.5772/65361

study is a 47-year-old healthy male volunteer, 174 cm of height and 78 kg of weight. GE's 64-slice CT scanner was used to take images of 500 × 500 mm (i.e., 512 × 512 pixels on the plane) cross-sections with 2.5 mm image-slice thickness. The images were taken from the extracranial skull base to the abdominal region. The geometries in-between slices (slice thickness: 1.25 mm) were created via an interpolation methodology. Compared with idealized geometries (see **Figure 3(a)** and **(b)**), the subject-specific geometry in **Figure 4** maintains more details of local airways and will have more realistic geometric influence on airflow and aerosol transport.

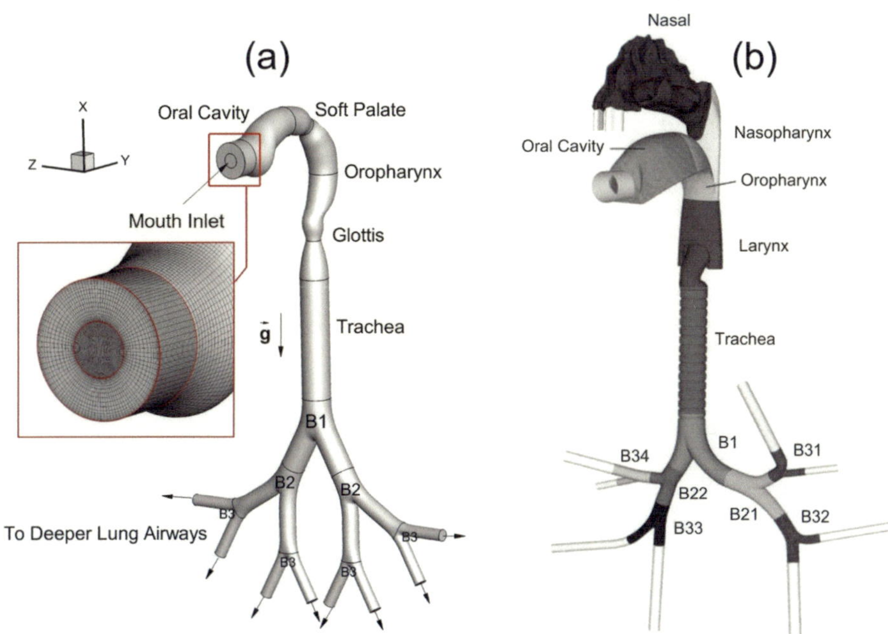

Figure 3. Configurations of two idealized human upper airway geometries: (a) symmetric idealized geometry from mouth to G3 with structured hexahedral finite-volume mesh and (b) asymmetric idealized geometry from mouth and nose to G3 (reprinted from [18]).

Creating high-quality boundary-layer mesh is paramount in computational mesh generation for the flow domain. It consists of multilayer hybrid tetrahedral/pentahedral elements near the wall surface to fully contain the viscous sublayers and to resolve any geometric features present there (see **Figures 3** and **4** as examples). Such high local mesh resolution can also help to accurately calculate values of near-wall derivatives, such as the deposition fluxes. Grid independence can be checked via indicators such as flow field solution and vapor deposition fraction (DF).

Realistic 3D imaging of the entire human respiratory system that is 100% accurate is currently unavailable because: (1) the lung consists of 2^{23} airways plus millions of alveoli, while the resolution of CT/MRI is insufficient to capture geometries in small scales, i.e., airways exceeding G9; (2) in-vivo measurements are difficult because the whole respiratory system geometry is time-dependent according to the respiratory movements.

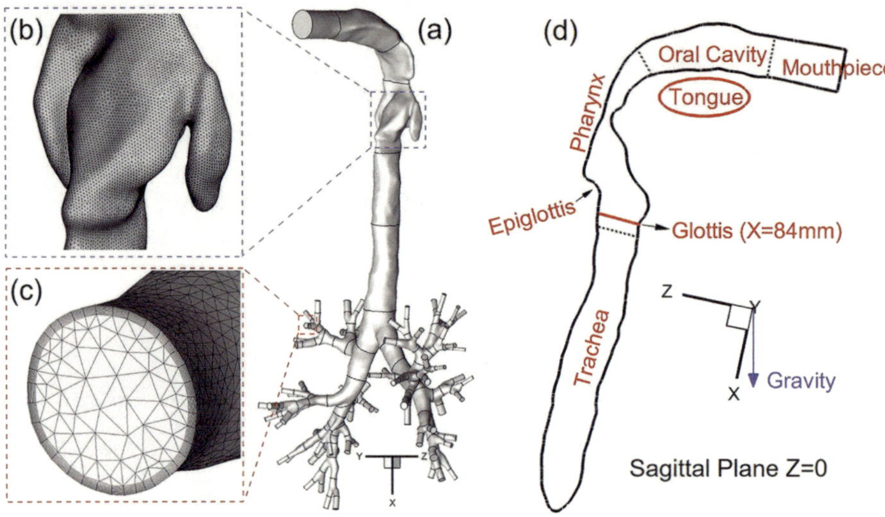

Figure 4. Configuration of a subject-specific human upper airway geometry from mouth to G9 with tetrahedral mesh including near-wall prism layers: (a) a subject-specific human upper airway model; (b) mesh details near the glottis; (c) mesh details at an outlet; (d) sagittal contour with regional names.

2.2. Governing equations

To provide a numerical description of airflow and aerosol behaviors, the governing equation system for CFPD modeling must be formulated. It contains physical and chemical conservation laws as well as supplementary equations (see **Figure 1**). The nonlinear differential governing equations subject to appropriate initial conditions and boundary conditions (BCs) will be discretized into algebraic equations and solved using different numerical techniques, e.g., finite difference or finite volume method. To promote the understanding of the physical and chemical characteristics of airflow and aerosol motions in lung, this section introduces the procedure to formulate equation systems for CFPD models for both conventional and unconventional aerosol transport and deposition.

2.2.1. *Airflow field equations*

The airflow dynamics in the respiratory tract is intrinsically unsteady, driven by the pressure differences under the action of cyclic breathing process. The incompressible Navier-Stokes (N-S) equation is employed to characterize airflow in the human respiratory tract, accompanied by continuity equation, energy equation, and constitutive equations [27], which can be written in tensor form as follows:

Continuity equation

$$\frac{\partial u_j}{\partial x_j} = 0 \tag{1}$$

in which u_j represents the fluid velocity and ρ is the air density.

Navier-Stokes (N-S) equation

$$\rho \frac{\partial u_i}{\partial t} + \rho \frac{\partial (u_i u_j)}{\partial x_j} = -\frac{\partial p}{\partial x_i} + \frac{\partial \tau_{ij}}{\partial x_j} + B_i \tag{2}$$

where p is the pressure, B_i is the body force including gravity, electromagnetic force, etc. The viscous stress tensor τ_{ij} in Eq. (2) is given by:

$$\tau_{ij} = \mu \left(\frac{\partial u_i}{\partial x_j} + \frac{\partial u_j}{\partial x_i} \right) \tag{3}$$

In Eq. (3), μ is the air viscosity.

Energy equation

$$\rho \frac{\partial \left(c_p T\right)}{\partial t} + \rho \frac{\partial (c_p \mu_j T)}{\partial x_j} = \frac{\partial}{\partial x_j} \left[k \frac{\partial T}{\partial x_j} \right] + \Phi + S_T \tag{4}$$

where k is thermal conductivity and c_p is specific heat capacity. Furthermore, S_T is the thermal sink or source term due to radiation, chemical reaction, phase change, etc. Φ is the dissipation function which can be given as:

$$\Phi = \tau_{ij} \frac{\partial u_i}{\partial x_j} \tag{5}$$

The equation of state as stated by the ideal gas law should complement the above equations for model closure, i.e., $pV_m = RT$, where V_m is the molar volume and R is the ideal gas constant.

At typical breathing rates, airflow through the oral airway region and first few generations is incipient turbulent. It becomes laminar at G6-8 and remains so thereafter. In order to capture the airflow structures in the laminar-to-turbulent flow regimes in human upper airways under common inhalation flow rates, the low-Reynolds-number (LRN) k-ω model and shear stress transport (SST) transition model are selected and adapted [27–32], based on their overall performance in predicting the onset of "laminar-to-turbulent" transition, their computational efficiency and reasonable accuracy as compared with large eddy simulations (LESs) [30]. For these Reynolds averaged Navier-Stokes (RANS) models, only averaged values of velocity, pressure, and other turbulence variables are calculated. The instantaneous fluctuation components can be recovered, being approximated by the eddy-interaction model (EIM) [33, 34].

2.2.2. *Particle dynamics equations (Lagrangian phase)*

For any given inlet concentration of effectively spherical particles (i.e., conventional aerosol particles), a Lagrangian frame of reference for the trajectory-computations can be employed. In light of the large particle-to-air density ratio and negligible thermophoretic forces, the generalized particle trajectory equation for particle *i* can be described as [35]:

$$\frac{d}{dt}(m_p \vec{u}_i^p) == \vec{F}_i^D + \vec{F}_i^L + \vec{F}_i^{BM} + \vec{F}_i^G + \vec{F}_i^{\text{interaction}} \tag{6}$$

Here, $\vec{u}_i^p$ and m_p are the velocity and mass of the particle, respectively; $\vec{F}_i^D$ represents the drag force which can be given by:

$$\vec{F}_i^D = \frac{1}{2}\rho C_D A_p (\vec{u}_i - \vec{u}_i^p) | \vec{u}_i - \vec{u}_i^p | \tag{7}$$

in which A_p is the projected particle area and C_D is the particle drag coefficient [3, 4, 10]. In Eq. (6), $\vec{F}_i^G$ is the gravity [35], $\vec{F}_i^{BM}$ is the Brownian motion-induced force [3, 10, 36], $\vec{F}_i^L$ is the Saffman lift force [37], and $\vec{F}_i^{\text{interaction}}$ is the force due to particle-particle interactions. The interactions may consist of van der Waals interaction and electrostatic interaction. The effect of the lubrication force, or near-wall drag modifications, may also need to be considered. All forces in Eq. (6) are point forces acting at the center of the particle, which employ specific empirical correlations corresponding to different particle aerodynamic properties.

2.2.3. *Advection-diffusion equation for vapor (Eulerian phase)*

For vapors and nanoparticles less than 50 nm in diameter, their inertial effects can be neglected. Thus, advection-diffusion equation can be used to describe vapor mass transfer which can be written as:

$$\frac{\partial Y}{\partial t} + \frac{\partial}{\partial x_j}(u_j Y) = \frac{\partial}{\partial x_j}\left[\left(\tilde{D}_a + \tilde{D}_{turbulence}\right)\frac{\partial Y}{\partial x_j}\right] + S_Y \tag{8}$$

where Y is the vapor mass fraction, $\tilde{D}_a$ is the vapor molecular diffusivity in air, $\tilde{D}_{turbulence}$ is the turbulence diffusivity which in laminar flow is zero, and S_Y is the source term.

http://dx.doi.org/10.5772/65361

2.2.4. Supplementary equations for modeling unconventional aerosol particles

The governing equations for conventional lung aerosol dynamics discussed above are for spherical particles with constant diameters. They are not sufficiently accurate to predict the transport and deposition of unconventional aerosol particles, i.e., nonspherical fiber and hygroscopic multicomponent droplets interacting with ambient vapors. Therefore, supplementary equations need to be introduced to encompass more physical and chemical mechanisms based on first principles [2–4, 10, 38].

2.2.4.1. Translational and rotational motion equations for ellipsoidal fiber

Based on the point-mass and point-force assumptions adapted by the conventional CFPD models, particles are in isotropic shape and their rotational motions are neglected. However, for nonspherical fibers with anisotropic shapes, the orientation determined by their rotational motions greatly influences the magnitudes and directions of the forces they experience in translational motion (see Eq. (6)). These forces will differ from those exerted on spherical particles with the same aerodynamic diameters. In order to capture this effect and provide accurate prediction of ellipsoidal fiber trajectory (see **Figure 5(a)** and **(b)**) based on realistic physical and chemical mechanisms, the conventional CFPD model was modified as follows [10, 13, 27]:

a. The translational equation (see Eq. (6)) was revised with force correlations specifically for ellipsoidal fibers;

b. Euler's rotation equations were introduced with the Euler quaternions to calculate fiber rotational motion and orientation at each time step;

c. The hydrodynamic torques acting on fibers are employed with correlations pertinent to ellipsoids.

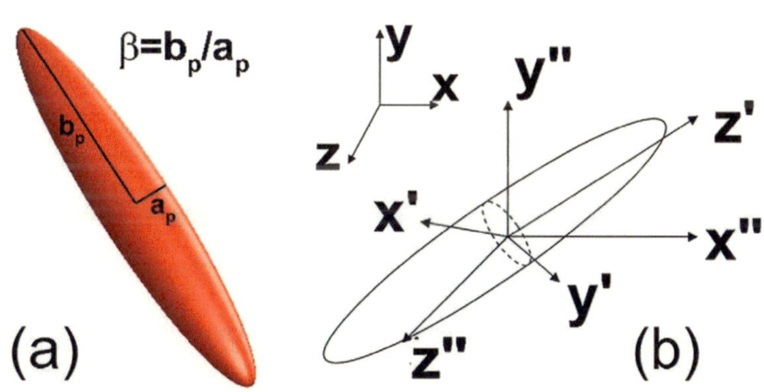

Figure 5. Sketches for ellipsoidal fiber modeling: (a) sketch of geometric parameters: semi-minor axis a_p, semi-major axis b_p, and aspect ratio β; (b) sketch of coordinate systems: space-fixed frame xyz, body-fixed frame $x'y'z'$, and co-moving frame $x''y''z''$.

The force and torque correlations for ellipsoidal fibers are documented in [10, 13]. Euler's rotation equations for ellipsoidal fiber orientation calculation in the body-fixed $x'y'z'$-frame (see **Figure 5(b)**) are given by:

$$I_{x'}\frac{d\omega_{x'}}{dt}-\omega_{y'}\omega_{z'}\left(I_{y'}-I_{z'}\right)=T_{x'} \tag{9a}$$

$$I_{y'}\frac{d\omega_{y'}}{dt}-\omega_{z'}\omega_{x'}\left(I_{z'}-I_{x'}\right)=T_{y'} \tag{9b}$$

$$I_{z'}\frac{d\omega_{z'}}{dt}-\omega_{z'}\omega_{y'}\left(I_{x'}-I_{y'}\right)=T_{z'} \tag{9c}$$

Here, $(I_{x'}, I_{y'}, I_{z'})$ are the particle moments of inertia about the principal axes x', y', z'; $(\omega_{x'}, \omega_{y'}, \omega_{z'})$ are the particle angular velocities with respect to the principal axes x', y', z'; and $(T_{x'}, T_{y'}, T_{z'})$ are the torques acting on the particle with respect to the principal axes x', y', z'. For ellipsoidal particles, $(I_{x'}, I_{y'}, I_{z'})$ can be written as:

$$I_{x'}=I_{y'}=\frac{\left(1+\beta^2\right)\cdot a_p^2}{5}m_p \tag{10a,b}$$

$$I_{z'}=\frac{2a_p^2}{5}m_p \tag{10c}$$

where a_p, b_p, and β are semi-minor axis, semi-major axis, and aspect ratio of the ellipsoidal fiber, respectively (see **Figure 5(a)**). Details of this advanced CFPD model, i.e., Euler-Lagrange method enhanced by Euler's rotation equation (EL-ER), are discussed by Feng and Kleinstreuer [10]. With different correlations of forces and torques acting on particles with specific anisotropic shapes, EL-ER method can be extended to simulate the transport and deposition of other nonspherical aerosol particles.

2.2.4.2. Governing equations for transport and phase change of droplet-vapor mixture

For liquid aerosols (i.e., droplet-vapor mixture), the phase change between droplet and vapor (see **Figure 6**) significantly affects their dispersion and deposition because the induced inertia changes during their transport. Therefore, the multicomponent mixture plus discrete-droplet (MCM-DD) model has been developed to simultaneously simulate multicomponent droplet-vapor interactions with evaporation and condensation effects during their transport in human respiratory systems [2, 3].

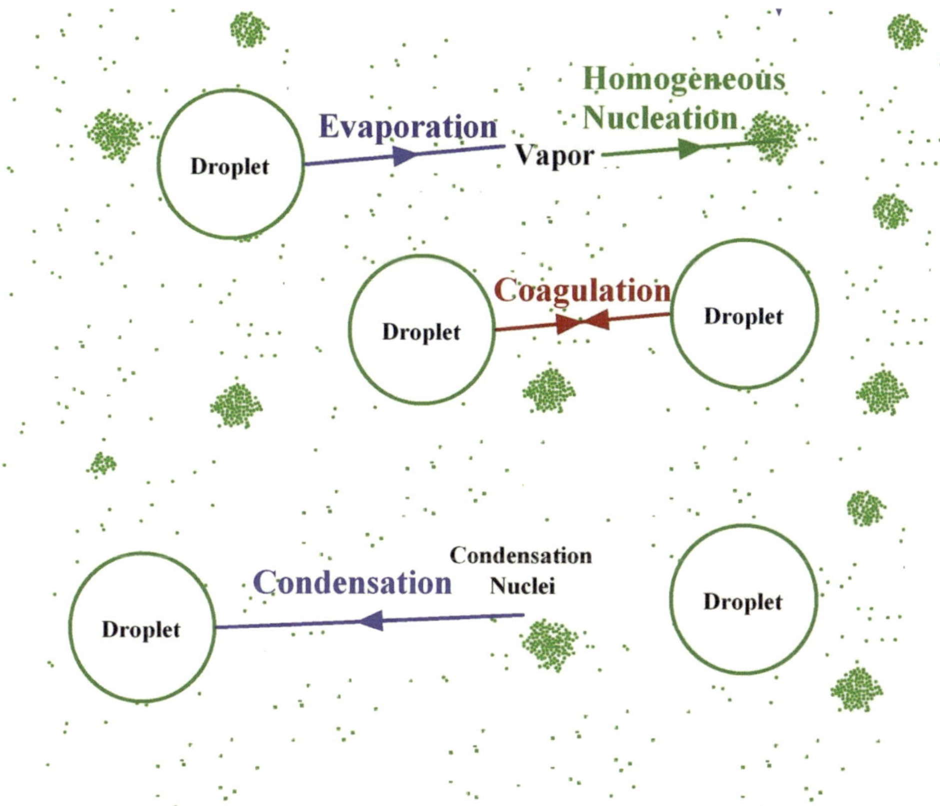

Figure 6. Schematic of droplet-vapor interactions and phase change dynamics.

2.2.4.3. Governing equations for continuous phase (air-vapor mixture)

Air-vapor mixture is described as a single continuous phase. Therefore, the conservation laws for mass, momentum, and energy are used to describe the air-vapor mixture transport phenomena and the advection-diffusion equation for vapor species transport (see Eqs. (1)–(5) and Eq. (8)). Droplet-vapor interaction will be realized by: (a) introducing source terms into the energy equation and vapor species transport equation (see Eqs. (4) and (8)) and (b) employing the local vapor mass fraction in the droplet mass conservation equation (see Eq. (15)). Specifically, the energy equation of air-vapor mixture can be stated as:

$$S_{v-d,s}^{(Y)} = \int_{t_{i,start}}^{t_{i,start}+\Delta t_f} \left(\sum_{i=1}^{N_{d,cell}} \left(\overline{j_s} A_d \right)_i \right) dt_d \,/\, \left(V_{cell} \Delta t_f \right) \tag{11}$$

where subscript $a–v$ means "air-vapor" and N represents the number of volatile components in droplets. The energy source term $S_{v-d}^{(E)}$ is due to the latent heat of evaporation or condensation which is released or absorbed by the droplets per local mesh cell:

$$S_{v-d}^{(E)} = \sum_{s=1}^{N} S_{v-d,s}^{(E)} = \left\{ \sum_{i=1}^{N_{d,cell}} \left[\left(\sum_{s=1}^{N} L_s \overline{j_s} \right) A_d \right]_i \right\} / V_{cell} \tag{12}$$

where $N_{d,cell}$ is the total droplet number in a specified mesh cell, subscript v–d means "between vapor and droplet", $\overline{j_s}$ is the average evaporation/condensation mass flux normal to the droplet surface, and L_s is the latent heat of liquid-vapor phase transition of the sth species.

Furthermore, the advection-diffusion equation of the sth vapor species can be given as:

$$\frac{\partial(\rho_{a-v} Y_{v,s})}{\partial t} + \frac{\partial}{\partial x_j}\left(\rho_{a-v} u_j Y_{v,s}\right) = \frac{\partial}{\partial x_j}\left[\rho_{a-v} \tilde{D}_{a-v,s} \frac{\partial Y_{v,s}}{\partial x_j}\right] + S_{v-d,s}^{(Y)} \tag{13}$$

Here, the local vaporized/condensed vapor mass flow rate of the aerosol components is added to its advection-diffusion equation as a source term $S_{v-d,s}^{(Y)}$ (kg m^{-3} s^{-1}) which is given by:

$$\frac{\partial\left(\rho c_p T\right)_{a-v}}{\partial t} + \frac{\partial\left(\rho c_p u_j T\right)_{a-v}}{\partial x_j} = \frac{\partial}{\partial x_j}\left[k_{a-v} \frac{\partial T_{a-v}}{\partial x_j}\right] + \Phi + \frac{\partial}{\partial x_j}\left[\sum_{s=1}^{N} h_s \rho_{a-v} \tilde{D}_{a-v} \frac{\partial Y_s}{\partial x_j}\right] + S_{v-d}^{(E)} \tag{14}$$

in which A_d is the droplet surface area. Additionally, dt_d represents the droplet phase time differential and Δt_f is the flow time-step length. It is worth mentioning that Eq. (14) describes the major physics of liquid-vapor mass change.

2.2.4.4. *Governing equations for multicomponent droplets*

The Lagrangian approach is chosen to track multicomponent droplets, neglecting rotational motion. The governing equations for discrete droplets are the translational equation (see Eq. (6)) as well as the mass and energy conservations of droplets and the supplementary equations. The mass conservation of droplets can be expressed as:

$$\frac{dm_d}{dt} = -\sum_{s=1}^{m} \int_{surf} j_s dA \approx -\sum_{s=1}^{m} (\overline{j_s} \cdot A_d) \tag{15}$$

where the average evaporation/condensation mass flux $\overline{j_s}$ is given by [39]:

$$\overline{j_s} = \rho_{a-v} C_m \left(Sh \cdot \tilde{D}_{a-v,s} / d_d\right) \ln\left[\left(1 - Y_{s,cell}\right) / \left(1 - Y_{s,vsurf}\right)\right] \tag{16}$$

in which $Y_{s,vsurf}$ and $Y_{s,cell}$ are the mass fractions of the sth vapor phase at the droplet surface and the cell center that the droplet is currently in. Eq. (16) assumes that the droplet radius is much smaller than the distance between the droplet mass center and the mesh cell center. Specifically, $Y_{s,cell}$ is determined by Eq. (13), which realizes the droplet-vapor interaction modeling.

The energy conservation of droplets is given by:

$$m_d c_{p,d}\left(dT_d / dt\right) = C_T\left(\frac{k_{a-v} \cdot Nu}{d_d}\right)\left(T_{a-v,cell} - T_d\right)A_d - \left(\sum_{s=1}^{N} L_s \overline{j_s}\right)A_d \tag{17}$$

where Nu is the Nusselt number.

2.2.5. *Lung deposited dose calculation for pharmacokinetic studies*

Different mechanisms can induce aerosol deposition in pulmonary route, i.e., diffusion, sedimentation, inertial impaction, interception, etc. Local deposition of particles in human airways can be quantified in terms of the deposition fraction (DF) or deposition efficiency (DE) in a specific region (e.g., oral airway, first, second, and third bifurcations). The DF is calculated as:

$$DF_i = \frac{\text{mass of aerosols deposited at region i}}{\text{total mass of aerosols inhaled}} \tag{18}$$

The DE can be calculated as:

$$DE_i = \frac{\text{mass of aerosols deposited at region } i}{\text{mass of aerosols entering region } i} \tag{19}$$

In addition to the above two traditional deposition parameters DE and DF, a deposition enhancement factor (DEF) can quantify local particle deposition patterns, which is defined as:

$$DEF = \left(DF_i / A_i\right) / \left(\sum_{i=1}^{n} DF_i / \sum_{i=1}^{n} A_i\right) \tag{20}$$

where A_i is the area of a local wall cell i, n is the number of wall cells in one specific airway region, and DF_i is the local deposition fraction in wall cell i. Clearly, high DEF-values suggest the presence of "hot spots" of local particle depositions as compared to the regional average. See Refs. [3, 10] on how to calculate depositions for nonspherical particles, droplets, and vapors.

3. Case studies

Employing the advanced CFPD models for unconventional aerosol particle dynamics discussed in Section 2.4, two case studies are briefly shown in this section: (a) the transport and deposition of ellipsoidal fiber as a new pulmonary drug carrier in a subject-specific human upper airway model and (b) the transport, hygroscopic behavior, and deposition of multi-component electronic cigarette (EC) droplets in an idealized triple bifurcation unit (TBU).

3.1. Transport and deposition of ellipsoidal fibers in a human upper airway

Due to the toxicity of airborne ellipsoidal fibers to human and hence the lack of experimental data using human volunteers because of the severe ethical constraints, it is essential to numerically investigate the transport and deposition of fibers in human respiratory systems. On the other hand, multifunctional micron fibers are also being used as pulmonary drug carriers for disease treatment. Therefore, the numerical investigations will provide data for estimations of dosimetry, safety, and the efficacy of drugs in the lungs which are also critical factors in the development of inhaled medicines.

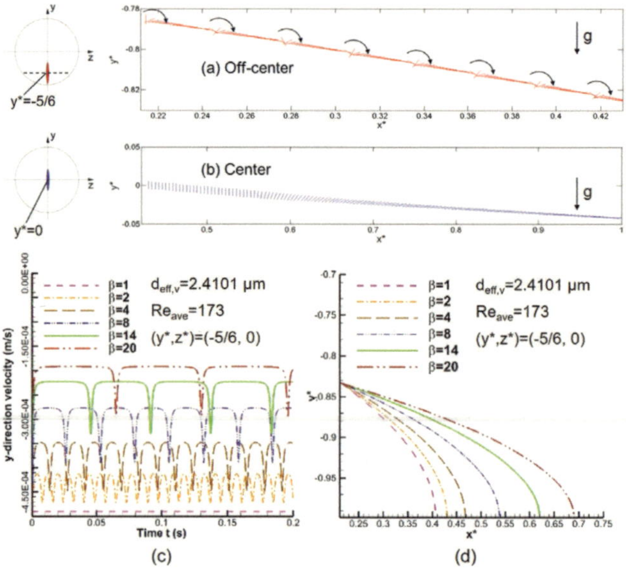

Figure 7. Ellipsoidal fiber translational and rotational motion in Poiseuille flow (Re_{ave} = 173) (reprinted from [10]): (a) ellipsoidal fiber long-axis directions released at an off-center position $y^* = -5/6$, $z^* = 0$; (b) ellipsoidal fiber long-axis directions released at center $y^* = 0$, $z^* = 0$; (c) y-direction velocities of fibers with the same volume and different aspect ratios; (d) trajectories for fibers with the same volume and different aspect ratios.

Considering the significant impact the anisotropic shape has on the accuracy of trajectory prediction, the one-way coupled EL-ER method discussed in Section 2.2.4.1 was employed [10] to calculate the translational and rotational motion of submicron ellipsoidal fibers of different aspect ratios ($1 < \beta < 20$). Brownian motion was also considered in both Poiseuille flow (see

http://dx.doi.org/10.5772/65361

Figure 7(a)–(d)) and complex flows inside a subject-specific human upper airway (see **Figure 8**).

To gain basic insight for ellipsoidal particle dynamics, Poiseuille flow with an averaged Reynolds number of $Re_{ave} = 173$ is considered where ellipsoidal fibers with randomly initialized incidence angles were released at different inlet-plane positions, their trajectories computed and visualized (see **Figure 7(a)–(d)**). **Figure 7(a)** and **(b)** show the rotational/translational motion of the ellipsoidal fibers ($\beta = 14$ and $a_p = 0.5$ μm) released at two locations, i.e., off-center and center. Released at an off-center position (see **Figure 7(a)**), the coupling of rotation and translation of the ellipsoidal particle can be clearly observed with periodical sudden turns of 180° in a "clockwise" manner. Released at the center (see **Figure 7(b)**), the flow velocity gradients were zero, i.e., no torques were affecting the particle. As a result, the particle kept its initial orientation till it reached a location where the velocity gradients and resulting torques were sufficiently large to rotate the particle. The frequency of the sudden rotation is influenced by the stability of the fibers at different locations of the flow field. It must be noted that using conventional CFPD models for spherical particles, the rotational motion is neglected and not able to be visualized.

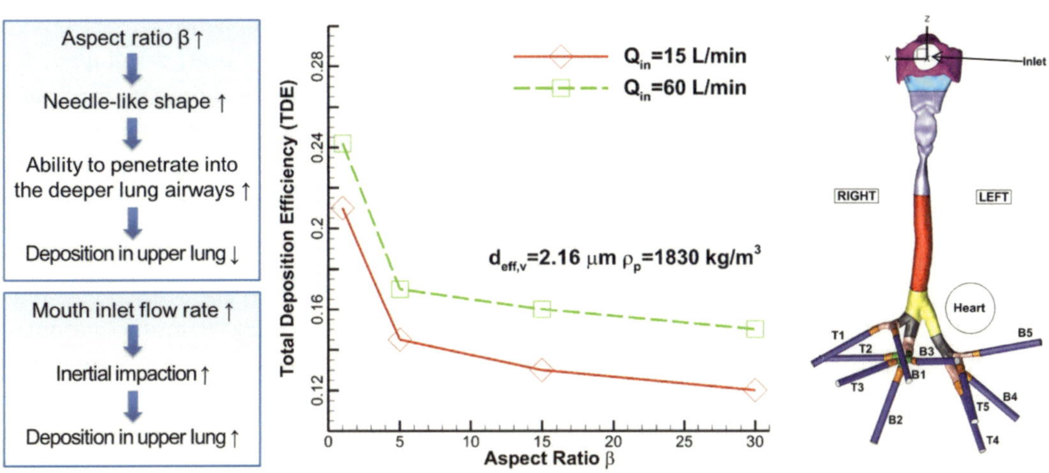

Figure 8. Total deposition efficiencies of ellipsoidal fibers, including spheres ($\beta = 1.0$, at two inhalation flow rates (Q_{in} = 15 L/min and 60 L/min) in a subject-specific human upper airway model as a function of aspect ratio.

Parametric sensitivity analyses were also performed for the aspect ratio effect on the translational and rotational motion for ellipsoidal fibers with the same equivalent volume diameter (see **Figure 7(c)** and **(d)**). Specifically, **Figure 7(c)** shows different y-direction velocities of fibers with the same volume but different aspect ratios. With the increase in aspect ratio β, the y-direction sedimentation velocity decreases and the rotation frequency of the fibers decreases. Indeed, with different aspect ratios, the Stokes resistance correction factors in the three principal directions of the ellipsoidal particle are all changing, which leads to the variation of the sedimentation velocity in y-direction. **Figure 7(d)** provides different trajectories of particles with the same volume and different aspect ratios. Particles with larger β-values have a stronger tendency to follow the mainstream and hence travel farther than a particle with a smaller β. Therefore, it can be conjectured that fibers are more likely to migrate deeper into the lung airways when compared to spherical particles of the same volume. In other words, when quasi-aligned to the flow, a fiber experiences a larger drag force in the gravitational direction (i.e., y-direction), so that particles with larger aspect ratios may not deposit in parallel flow.

To further investigate the transport characteristics of ellipsoidal fibers with different aspect ratios and the same volume, parametric studies on fiber transport in a subject-specific human upper airway with 20-mm inlet are performed. Specifically, a comparison of total deposition efficiencies of spheres and ellipsoidal fibers for two inhalation flow rates is shown in **Figure 8**. The total deposition data indicate that slender toxic ellipsoidal fibers (i.e., those with higher aspect ratios) are potentially more harmful than thicker ones due to their ability to penetrate into deeper lung regions when somewhat aligned with the mainstream flow. In contrast, to treat deeper lung diseases using pulmonary drugs, slender ellipsoidal fibers as drug carriers may enhance the drug delivery efficacy compared to conventional spherical ones.

3.2. Size change dynamics of multicomponent droplets in a G3-G6 TBU

As tobacco-smoking is addictive with potentially detrimental effects and hence related to high health-care cost, alternatives such as electronic cigarettes (ECs) have been developed. EC aerosols consist of multicomponent droplets and vapors, e.g., water, nicotine, glycerol, and PG. Although ECs can look like and ultimately should "taste" like conventional cigarettes, their design and function as well as product quality are very different. Debates are still going on for the use of ECs, with reported positive and negative aspects of EC consumptions [4]. Therefore, investigations on the differences between conventional cigarette smoke and EC aerosols are of great interest for EC product design as well as government regulations.

In this study, as the first step, the focus is on hygroscopic growth of nanosize multicomponent droplets and droplet-vapor interactions during transport with subsequent deposition in a triple bifurcating unit (TBU) representing G3–G6 lung airway (see **Figure 9**). Using the MCM-DD model derived in Section 2.2.4.2, electronic cigarette (EC) droplets with different compositions were compared with conventional smoke particles (CSPs) as well as solid particles on their different behaviors of evaporation/condensation dynamics during their transport and deposition in the TBU, with body temperature T = 310.15 K and relative humidity RH = 99.5% [2].

Figure 10(a) shows the comparisons of hygroscopic behavior between three EC droplets with different compositions and two conventional cigarette smoke particles (CSPs) with different compositions (see **Table 1** for compositions). At RH = 99.5%, both EC-droplets and CSPs grow due to condensation effect. However, CSPs #1 and #2 grow less than EC-droplets #B1 to #B3. This suggests that with initial droplet diameter and ambient environment (humidity) being the same, EC-droplets #B1 to #B3 will be always larger in size than the liquid CSPs #1 and #2. Further, these EC-droplets with larger diameters with have higher inertia, and thus being less affected by the local airflow structure than CSPs.

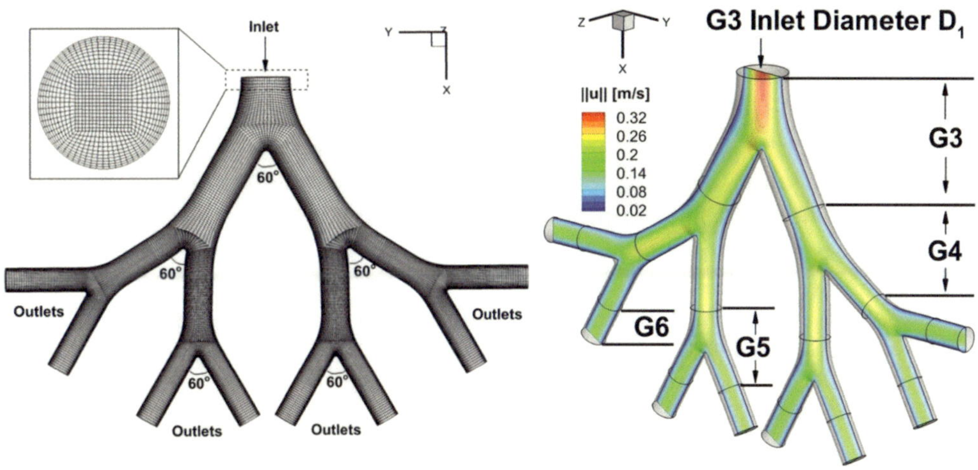

Figure 9. G3–G6 triple bifurcation unit (TBU) with hexahedral mesh and airflow velocity contour for Re_{in} = 61.

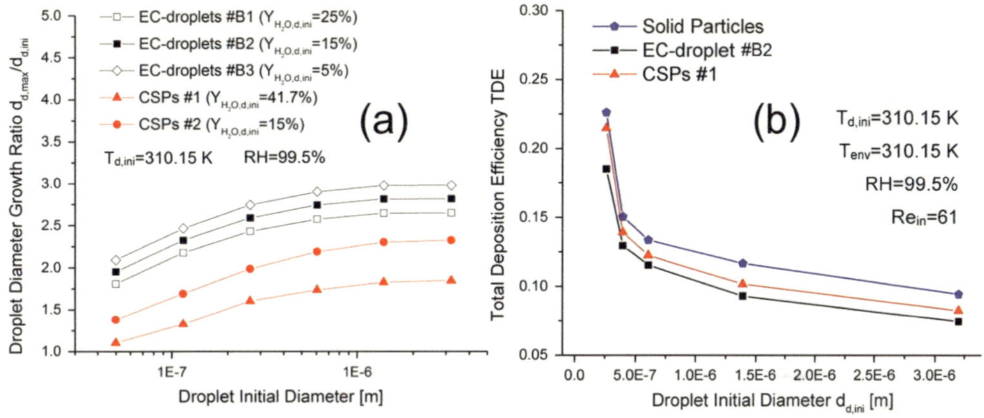

Figure 10. (a) Droplet diameter growth ratio comparisons between CSPs and EC-droplets; (b) total deposition efficiency of EC-droplet #B2, CSPs #1 and solid particles with different initial diameters (reprinted from [2]).

	Water	Glycerol	PG	Nicotine
EC Droplet #B1	0.25	0.398	0.332	0.02
EC Droplet #B2	0.15	0.498	0.332	0.02
EC Droplet #B3	0.05	0.598	0.332	0.02

Table 1. Composition of different EC droplets (mass fraction) used for modeling.

Figure 10(b) shows the total deposition of EC-droplets #B2, CSPs and solid particles in the G3-G6 TBU, assuming the same initial particle diameters (i.e., 264 nm $\leq d_{d,ini} \leq$ 3200 nm). It can be observed that with the same initial diameter, the total deposition fraction for the three particles is EC-droplets #B2 < CSPs #1 < solid particles. Clearly, the hygroscopic growth of droplets and CSPs limited their Brownian diffusion, thereby reducing TDE in G3–G6. Since this effect is more prominent for EC-droplets than for CSPs (see **Figure 10**), the latter therefore has higher total deposition efficiency values. In other words, CSPs #1 deposit more in the G3–G6 TBU than #B2 EC-droplets due to lower hygroscopic growth and stronger Brownian diffusion effect. It can thus be conjectured that EC-droplets may deposit more than CSPs #1 up to the first bifurcation, due to more pronounced inertial impaction. In contrast, those EC-droplets tend to deposit more after the first bifurcation in G3–G6 TBUs and deeper lung regions where Brownian diffusion dominates deposition. Results also indicate the important effect of hygroscopic behaviors of droplets on the accuracy of deposited-dose predictions.

4. Challenges and future directions

Challenges faced by the current CFPD models and numerical studies for unconventional lung aerosol dynamics are mostly induced by the point-mass and point-force assumptions, the resolution of medical imaging techniques, the computational hardware capabilities, as well as the gap between CFPD prediction capabilities and the needs of toxicologists and health-care providers. Specifically,

1. The point-mass and point-force assumptions raise the difficulty to apply CFPD modeling techniques on aerosol transport and deposition simulation in the alveolar region, where the particle size will not be much smaller than a single mesh element and the assumptions are not valid.
2. The current resolution of medical imaging techniques is not sufficiently high to identify small lung airways beyond G8. Therefore, most CFPD models are still for human upper airways, and it is impossible to reconstruct a complete subject-specific human respiratory system using CT/MRI imaging data. This restricts CFPD models from simulating aerosol transport in a full inspiratory-expiratory breathing cycle, i.e., the quantitative prediction of exhalation clearance of aerosols.
3. Due to the computational resources limitation, most CFPD models still use simplified mechanisms to reach good compromises between accuracy and computational efficiency, e.g., one-way coupling for dilute particle suspensions. Therefore, CFPD models are not suitable for dense particle suspensions in which the particle-particle interactions are not negligible.
4. Although CFPD models are able to provide high-resolution lung deposition patterns for unconventional aerosols, the deposited dose cannot be used directly for health risk assessment or drug delivery efficacy evaluation. Improvements need to be made for the current CFPD models to calculate the tissue and delivered doses.

This section shows the roadmap to bridge the current simulation techniques to emerging toxicological/pharmaceutical paradigms to overcome the limitations of current numerical simulations. The long-term goal is to build an accurate and cost-effective digital respiratory simulation framework of subject-specific exposure to inhalable aerosols under realistic breathing cycles. A proposed framework and kinetic process of this multiscale model, i.e., a hybrid CFPD-LBM-PBPK model (see Sections 4.1–4.3), is shown in **Figure 11**. Details of future researches for extensions and improvements of the current CFPD model and the development of the next-generation model are discussed below.

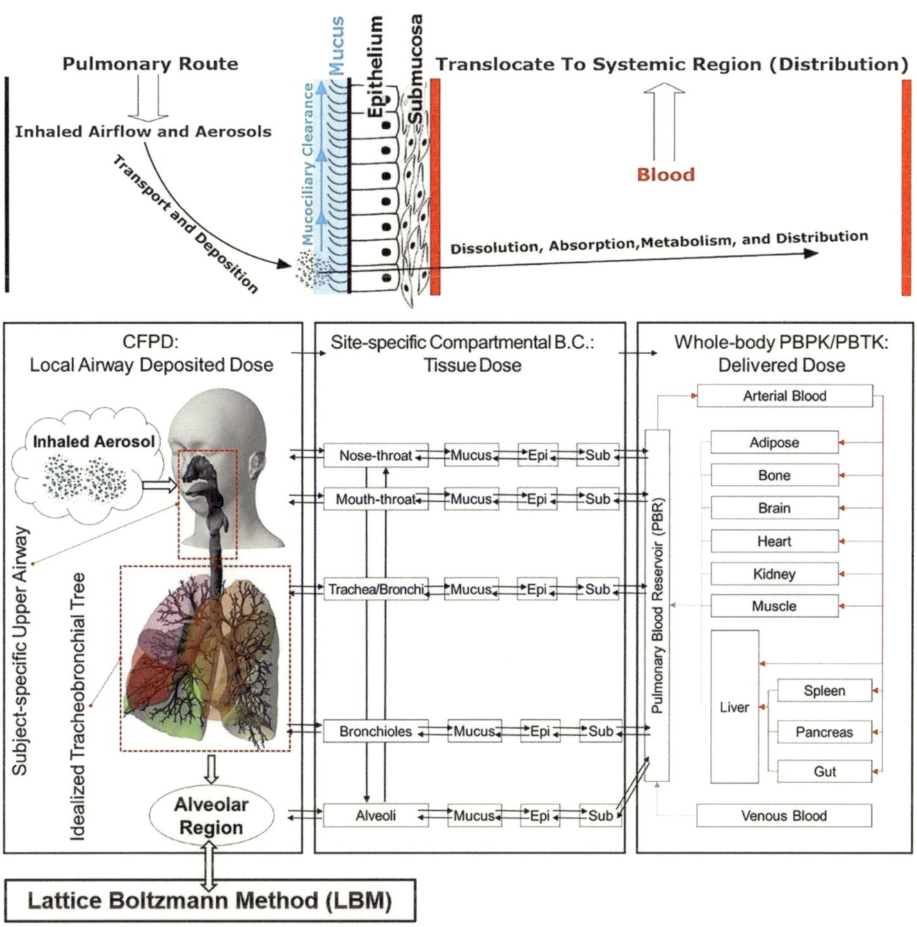

Figure 11. Schematic diagrams of the proposed multiscale CFPD-LBM-PBPK model: model framework with a complete human respiratory system geometry and kinetic processes after aerosol inhalation.

4.1. Reconstructing and modeling the entire conducting zone of human lung

In order to provide full-scale regional lung deposited doses in the oral/nasal cavity, conducting and respiratory zone, it is necessary to find methodologies to represent or reconstruct the complete human respiratory system geometry, including the smaller airways, which are not visible via CT/MRI imaging. Accordingly, two methods can be further developed: (1) coupling methods with single-path or multiple-path lung airway trees and (2) virtual lung model with idealized airways for higher generations covering the entire lung conducting zone (G0–G17). This is crucial to create a full-scale lung aerosol dynamics simulation featuring full breathing cycles.

4.1.1. *Coupling methods with single-path or multiple-path lung airway trees*

Coupling methods combine coupled boundary conditions with truncated lung airway trees to represent the whole conducting zone. Realization of coupling methods can be performed in the following steps:

1. Divide the whole lung airway trees into segments [40] or truncate the airway tree with single path or multiple paths left [41–43];
2. Simulate each segment in series or in parallel;
3. The boundary conditions applied to each truncated airway ends will be obtained from the previous simulation of the next larger section (inhalation) or the next smaller section (exhalation).

Kleinstreuer and Zhang [40] used the TBU of Weibel type A and constructed the symmetric and in-plane tracheobronchial airways from G0 to G15, and used coupling method to simulate airflow and particle deposition in the lung airways with steady-state inhalation flow rate Q_{in} = 30 L/min. Walters and Luke [42] created an 8-generation airway tree with random connection angles following Weibel's lung morphology. They applied coupled pressure boundary conditions for inhalation cases, and validated the coupling method for inhalation cases. They extended the application of this coupling method on particle transport in lung airways [43]. Tian et al. [41] developed the stochastic individual path (SIP) and multiple stochastic individual path (MSIP) approaches to represent the whole-lung geometry. Compared to the coupling method applied on detached airway geometries [41], this model is able to simulate transient breathing scenarios. Recently, similar models have also been developed [44, 45].

Since modeling the whole lung is computationally expensive, using the coupling method is potentially advantageous for fast-solution purposes. However, accuracy of the coupling methods still needs to be tested compared with using virtual lung model containing the entire conducting zone without any truncations.

4.1.2. *Virtual lung model for the entire conducting zone*

To encompass the entire conducting zone, the main objective of virtual lung model reconstruction is to develop a stochastic algorithm that generates 3D asymmetric human lower lung airways, which can then be connected to the subject-specific upper lung airways [46–48]. To establish such a virtual lung model, the upper airway (i.e., mouth and nasal airways, throat, and trachea) will be exactly reconstructed using image processing software that converts CT/MRI images. Conducting airways, from the trachea down to the level of the terminal bronchioles, will be generated via a stochastic algorithm [46–48]. Specifically, by generating

http://dx.doi.org/10.5772/65361

geometric variables (i.e., airway diameters, branching angles, and rotation angles of successive branching planes) stochastically with morphometric restrictions [46], following fractal model (i.e., Hess-Murray law), or grow from 1-D centerline airway tree [48], a complete tracheobronchial airway tree starting from the trachea can be constructed. The accuracy of the geometry can be judged in two ways: (1) the airways follow the centerlines of a 1D lung airway tree and (2) it conforms to surface data of the airways segmented from medical imaging. The combined geometry can be exported as a stereolithography (STL) file that can be used for mesh generation as needed for computer modeling and simulations.

Compared with coupling methods, constructing a virtual lung model will enable CFPD simulations with more realistic boundary conditions. It will be computationally achievable to simulate the multiphase inhaled aerosols in the entire lung conducting zone, with foreseeable increase in computational power.

4.2. Lattice Boltzmann method (LBM) for aerosol transport in alveoli

A portion of the inhaled aerosol will enter the respiratory zone (G17–G23) of human lung, which is mainly the alveolar region. In this case, the difficulty in using CFPD model is that the point-mass and point-force assumptions do not hold and particle-particle interactions are not negligible. Lattice Boltzmann method (LBM) can be employed for the two-way fluid-particle interaction and many-particle interaction in alveoli, due to its advantage in resolving particles that are comparative or larger than the mesh cells in size.

4.2.1. Background and literature review

Lattice Boltzmann method (LBM) solves the Lattice Boltzmann equation (LBE) to obtain the probability density distribution, which is then used to obtain the pressure and velocity field through integration. Details of the LBE, discrete speeds, discretization, and boundary conditions are given in [49] and avoided here for brevity. LBM consists of two main steps: (1) propagation: particle distribution moves along the lattice bonds to neighboring lattice nodes and (2) collision: particles on the same node interact and adjust velocities to conserve momentum and mass. Communication between nodes in the lattice is required only in the propagation step requiring message passing or update of the shared memory location.

LBM has been applied to the airflow and particle transport simulations in the human airways due to the advantages mentioned above compared with CFPD models. Many studies focused on the flow structures in nasal cavities. For example, Finck et al. [50] first employed LBM for steady simulations of laminar nasal flow. The model predictions agree well with those of finite-volume based model using structured grids. However, the LBM has high flexibility in fast grid generation and easy boundary-condition implementation for such complex geometries compared to conventional Navier-Stokes solvers. Hörschler et al. [51] compared steady and unsteady flows in real nasal cavity, and concluded that the transient effects are only important at $Re < 1500$. Henn et al. [52] coupled Lagrangian particle tracking method to LBM to simulate aerosol dynamics in a human nasal cavity. They concluded that transient simulations are necessary to study particle dynamics in nasal cavities. Other studies modeled airflow in the nasal cavities with diagnosed pathologies [53].

There are also existing researches focusing on air flow simulation in human upper airway using LBM. Ball et al. [54] showed that LBM yields higher fidelity than the RANS models for laminar-to-turbulent transition flow regime predictions inside an idealized human extrathoracic airway, as compared to experimental results. Schroeder et al. [55] presented LB simulation results of pulsatile flow at two breathing frequencies in a human respiratory system down to G6. They showed that inspiratory flow is characterized by flow separation and secondary flow, while expiratory flow is characterized by more homogeneous structure but higher vorticity. This is in contrast with the study by Eite et al. [56], which showed that less vorticity is generated at exhalation than at inhalation in a nasal cavity. Krause [53] investigated respirations in the upper part of the human lungs using LBM coupled with a model for the rest of lung, thereby constructing a two-scale whole lung model. This latter model assumed that the lower part of the bronchial tree can be described by a symmetrical dyadic tree of pipes, where Poiseuille type flow holds. Imai et al. [57] developed a LBM with GPU acceleration using an adaptive meshing method. The model was used to study the effect of breath holding on the deposition of microparticles in pulmonary airways during inhalation. The subject-specific airway model contains thirteen generations of airways. The study found that breath holding can effectively increase particle deposition; yet the effect varies for different particle sizes. Wang and Elghobashi [58] analyzed flow structures in human upper airways including the nasal cavity, the pharynx, the larynx, and the upper part of the trachea for both healthy and diseased subjects. The laminar-transitional-turbulent flow in the airway was captured in inspiration, and laminar flow was found during expiration. The authors proposed that the maximum positive value of the local time-averaged second derivative of pressure can be used to locate the obstructions due to disease in the upper airway. It should be noted that most studies claimed that LBM provides better accuracy for air flow simulation in respiratory systems. However, Chen and Gutmark [59] did a comparison study of airflow in an idealized human extrathoracic airway. They claimed that LES is more accurate in predicting root-mean-square flow field compared with Reynolds stress model and LBM.

In contrast to many existing LBM simulations on airflow patterns in human respiratory systems, there are only a few studies using LBM for lung aerosol dynamics. Henn et al. [52] used LBM coupled with Lagrangian tracking method to study microparticle transport in human nasal cavities. Trunk et al. [60] developed an Euler-Euler LBM method to investigate the particle dynamics in an idealized bifurcation geometry representing the split from trachea to main bronchi. Particle capturing BC and numerical stabilization were discussed for lung aerosol dynamics simulations. Li and Kleinstreuer [61] analyzed transient airflow in two-dimensional alveoli and bifurcating alveolar ducts. The results highlighted the importance of alveolar geometries and shapes, as well as the effect of the moving alveolus walls. Implications of these effects on particle deposition in the alveolar sac were discussed. However, simulation of particle suspension has not been performed.

4.2.2. LBM vs. CFPD models

As compared with the CFPD models, LBM has advantages in the following aspects [54]: the convection operator is linear in LBM; the conservation laws are exactly preserved in LBM; the calculations are arithmetic operations in LBM; and the boundary conditions (BCs) like no-slip or moving boundaries are easy to implement. In addition, the entropic lattice Boltzmann method is attractive in modeling transitional and fully turbulent flows for high-Re simulations [62]. It should be noted that when the particle size is much larger than the lattice size, LBM can readily account for the two-way fluid-particle interaction and many body interaction by accurately calculating the hydrodynamic force and torque exerted on the particle by the fluids [63–65]. Also, Brownian motion of particles arises spontaneously from stochastic fluctuations in the fluid stress tensor, avoiding application of a random force on particles. Therefore, LBM is suitable for the simulation of aerosol dynamics, especially in the alveoli sacs. Furthermore, LBM can also be used to simulate the mucociliary clearance process when coupled with immersed boundary method or finite difference method [66, 67].

In conclusion, LBM is a rapidly developing method in lung aerosol dynamics simulation. Many researches have been done concerning airflow dynamics. On the other hand, LBM simulation including particle transport and deposition in alveoli is still largely unexplored. Much more efforts are demanded to couple particle transport with airflow simulation using LBM, with both sub-grid-size particles, and macro-size particles.

4.3. Physiologically based pharmacokinetic (PBPK) models for aerosol translocation into systemic regions

4.3.1. Background and literature review

Carried by ultrafine aerosols, toxicants and therapeutic components can easily translocate from lung deposition sites into the systemic circulation, and accumulate in extrapulmonary organs. Therefore, after obtaining deposited dose of environmental pollutants or pulmonary drugs in the human respiratory systems, the judgment of potential adverse effects or curing effectiveness depends on agents' toxicokinetic/pharmacokinetic information with more insights into their in-vivo behavior, and then toxicology, i.e., absorption, distribution, metabolism, and elimination. In order to provide comprehensive regional data for risk assessment, it is necessary to link a whole-body PBPK model to CFPD models, which extends exposure and deposition modeling to health endpoints. The integration of preclinical data generated by CFPD-PBPK models is also a key to optimize the drug delivery processes for clinical practice.

Physiologically based pharmacokinetic (PBPK) models are based on differential equations that are designed to simulate the absorption, distribution, metabolism, and excretion (ADME) in compartments which are connected biological regions, usually in whole animals or humans [11, 66]. Realistic descriptions of physiology and relevant pathways of metabolism can be incorporated into PBPK models. The history and details of PBPK models have been well discussed in [67, 68].

Focusing on the development of hybrid CFPD-PBPK models, the respiratory system serves as one compartment, but has significant influence on the whole body as it is the main barrier between inhaled aerosols and systemic circulation (see **Figure 11**). As direct inputs to the PBPK model, regional airway-tissue doses must be accurately calculated from the lung deposited doses acquired from the CFPD simulations. Therefore, the aim of incorporating CFPD models with PBPK models is to develop a new boundary condition for site-specific airway-tissue dosimetry for PBPK modeling that includes essential mechanisms, such as clearance, tissue reaction, tissue metabolism, diffusion, and blood perfusion. Assuming a linear first-order constitutive relationship, the material conservation law for compartment *i* of the airway tissue can be expressed as:

$$\partial c_i / \partial t = \sum_{j=1}^{N} (k_{ij} c_j - k_{ji} c_i) + S_i(t) \tag{21}$$

where c_i is the amount of pathogen per unit volume accumulating in compartment *i*, k_{ij} is the transfer rate from compartment *j* to *i*, and S_{ij} is a general source term representing exterior inputs, reactions, clearance, etc. The transfer rate, k_{ij}, can be determined from existing experimental data, or in vivo deposition/retention and mass transfer data obtained from laboratory animals. Other biological material properties and site-specific boundary conditions for the PBPK model will be obtained from the open literature.

Efforts have been made to link CFPD and PBPK/PBTK models for simulations of inhaled drugs or toxicants, from deposition to systemic regions of interests [69–75]. For example, Rigg et al. [69] established a numerical model for particle dissolution, absorption, and clearance (DAC) in the nasal cavity. Their DAC model can estimate the tissue dose for particles after deposition, which are the more accurate inputs than deposited dose for PBPK/PBTK calculation. Using subject-specific respiratory system geometries of different species, Corley et al. [70] compared the parameter difference between humans and animals and adapted the nasal CFD-PBPK model [71] to perform interspecies variability analyses and sensitivity analyses for vapor exposure. Anatomic, physiological, and biochemical parameter values were also listed [70]. Parameters such as airway geometry, inhalation flow rates, vapor concentration, partition coefficient between air and tissue, tissue thickness, and metabolism rate were identified as the significant factors that influence local vapor uptakes. Recently, they extended the CFD-PBPK model to transient simulation for toxic vapors [72]. Furthermore, PBPK models with different compartmental divisions have been established to estimate pulmonary drug delivery [73, 74]. Parameter values were also provided in the context.

4.3.2. Comparisons among CFPD-PBPK, CFPD, and PBPK models

Compared with PBPK models, CFPD-PBPK models have the advantage in considering the subject variabilities induced by interindividual morphologies of human respiratory systems, which can be reflected by the different local deposition patterns provided by CFPD calculations. Additionally, CFPD-PBPK models extend exposure and deposition modeling of CFPD to health endpoints, i.e., delivered dosage estimations at site of interests in the whole human body.

Although research efforts have been made to develop hybrid CFPD-PBPK models as discussed in Section 4.3.1, none of the existing models display the realism needed for accurate dosimetry calculations, because they ignored the influence of essential physical and chemical mechanisms such as hygroscopic characteristics on the dispersion and deposition of droplets in the respiratory system, and used incomplete human lung airway geometries.

Therefore, future directions to develop more accurate and computationally efficient CFPD-PBPK models are to make more realistic compartment divisions, mechanism selections, and parameter value determinations to reflect the underlying aerosol dynamics in systemic regions [76]. For example, the compartmental division of cellular-level PBPK model should be based on specific anatomy and mass transfer characteristics of different lung regions. It is also important to collaborate closely with experimentalists and clinical doctors to find reliable data sets reflecting responses and pharmacokinetics for evaluating the performance of the CFPD-PBPK model [77]. Moreover, the determination of human physiological and biochemical parameters requires strong collaborations. These parameters include breathing rate, skin surface area, cardiac output, tissue blood flow rates, tissue volumes, partition coefficients, and biochemical rate constants [68]. Subject-variability must be considered too, i.e., physiological and biochemical parameter differences due to the differences in interindividual response to the biological effects of inhaled aerosols. It will not only contribute to the evaluation of the confidence in PBPK models on the lack of precise knowledge of the parameter values, but also contribute to the uncertainty analyses for the extrapolations between animals and humans. The error source must be identified and reduced to enhance the accuracy of the numerical results [75].

4.4. Others

Other numerical methods that can be integrated or employed to improve or replace current CFPD models are direct numerical simulations (DNSs) for turbulence [78], complete numerical simulations (CNSs) for multiphase flow [79], discrete element methods (DEMs) [80], etc. They will be able to extend the CFPD model capabilities by encompassing more underlying physics (e.g., particle-particle interactions) and less simplifications (e.g., point-mass and point-force assumptions).

4.4.1. Complete numerical simulations (CNSs) for multiphase flow

Complete numerical simulation (CNS) methods are numerical techniques where the Navier-Stokes equations are applied to finite-size particles instead of introducing point-mass and point-force assumptions. Particle motion is governed by the Newton's second law, based on the hydrodynamic forces calculated from the fluid equations. For CNS-based multiphase flow, all surface and collision forces due to fluid-particle and particle-particle interactions should be accurately calculated to obtain the velocity, pressure, and stress fields surrounding each particle [35]. Specifically, mesh scales for CNS methods are much lower than particle sizes;

hence, surface forces such as drag force and lift force, as well as interaction force between particles can be directly integrated from the shear stress and pressure distribution along the particle surfaces. This avoids empirical correlations, such as drag and lift coefficients.

Compared with CFPD models which are based on Euler-Euler or Euler-Lagrange methods, CNS is a mesoscale model that can accurately describe motions of particles with arbitrary shapes. As such, CNS of multiphase flows can promote fundamental understanding of the underlying physics, e.g., the nonlinear and geometrically complicated phenomena of particle-particle and particle-wall interactions.

With the enhancement of the computational power in the future, CFPD models will be replaced by DNS and CNS methods to simulate unconventional lung aerosol dynamics. However, the computational cost of DNS and CNS methods is still too high for engineering application. Applying these methods to a large number of particles is not realistic according to the current computational resources [79].

4.4.2. Discrete element method (DEM)

Discrete element method (DEM) can provide transient forces and torques acting on individual particles. DEM was first proposed by Cundall and Strack [81] based on molecular dynamics. The most attractive feature of DEM method is its relatively efficient algorithms for the contact detection and contact force calculation between arbitrary shaped particles [82]. Soft-sphere models are the most common type of models used in DEM, which simplify particle-particle interaction force with spring-dashpot-slider systems [27]. Determinations of the stiffness coefficients and damping coefficients are essential in DEM interaction models. Contact forces are calculated in two steps in DEM methods [83]:

1. Contact detection: identify which particles are contacting;

2. Contact resolution: the actual contact forces are calculated. This step involves the accurate calculation of the contact geometry and kinematics, typically characterized by the overlap depth/separation and sometimes also relative tangential motion.

Additionally, for nonspherical particles, representation of the particle shape is also essential in DEM simulations. Two major treatments of the particle shape representation are widely used for the contact detection and contact force calculation between nonspherical particles [84]: superquadric method and multisphere method. The algorithm to integrate DEM with Euler-Euler and Euler-Lagrange models (see **Figure 12** for a standard flow chart) for dense particulate flow is straightforward [38]. However, there are two key challenges:

1. Spring-dashpot-slider systems are not designed for micro-/nano-particles. Coefficients, i.e., equivalent stiffness and damping coefficient, need to be determined via experiments specifically for ultrafine particles. Or else, the validity of using soft-sphere models needs to be tested.

2. DEM is not computationally efficient on handling aerosol transport and deposition in human respiratory systems, in which the difference in characteristic dimensions between

particles and flow domains is too large. Therefore, efficient algorithms need to be developed.

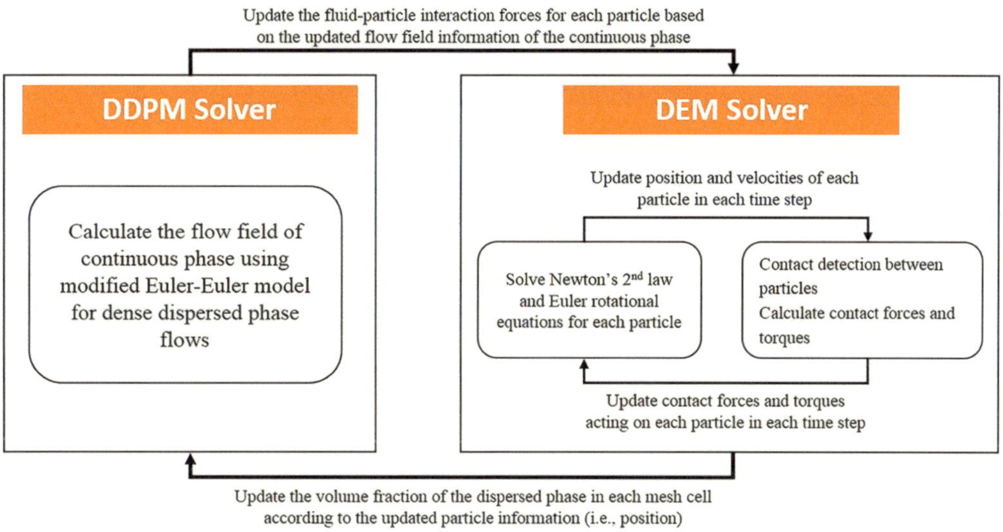

Figure 12. The algorithm of coupling DEM with conventional multiphase flow models.

5. Summary

In this chapter, CFPD models have been systematically overviewed, with emphases on simulating the transport and deposition of unconventional ultrafine aerosol particles in human upper airways. Compared to other numerical models for inhaled aerosol dose [1], CFPD models provide more detailed characterizations of airflow and aerosol deposition patterns with a reasonable degree of accuracy and computational cost.

However, in order to provide comprehensive tissue and delivered dosage data for risk assessment or drug delivery efficacy evaluation, it is necessary to improve current CFPD models and extend lung deposition calculation by integrating with other simulation techniques to build a multiscale dosimetry model (see **Figure 11** for the framework). This multidisciplinary effort requires expertise in medical imaging, airway geometric reconstruction, computational techniques, pulmonary physiology and medicine, fluid mechanics, etc. From a physics viewpoint, the future numerical models for lung aerosol dynamics will provide comprehensive analysis of the fate of inhaled particulate matter under realistic conditions, namely, actual particle characteristics, inhalation flow rates, and human respiratory systems. The anticipated results will greatly increase understanding of distinct particle transport/deposition phenomena under realistic inhalation conditions as well as quantify the knowledge base for dosimetry and health-effect studies in the case of toxic particles, and optimal drug-aerosol targeting in the case of therapeutic particles. Relevant numerical researches will lead to understanding and quantification of dosimetry effects of inhalation hazards, more accurate evaluation of toxic particle fate in the human lung, and improved estimates of drug delivery points for treating lung and other diseases.

Author details

Yu Feng[1*], Zelin Xu[2] and Ahmadreza Haghnegahdar[1]

*Address all correspondence to: yu.feng@okstate.edu

1 School of Chemical Engineering, Oklahoma State University, Stillwater, OK, USA

2 Department of Mechanical Engineering, North Carolina State University, Raleigh, NC, USA

References

[1] Dong J., Shang Y., Inthavong K., Tu J., Chen R., Bai R. Comparative numerical modelling of inhaled nanoparticle deposition in human and rat nasal cavities. Toxicol. Sci. (in press). DOI: 10.1093/toxsci/kfw087.

[2] Feng Y., Kleinstreuer C., Rostami A. Evaporation and condensation of multicomponent electronic cigarette droplets and conventional cigarette smoke particles in an idealized G3–G6 triple bifurcating unit. J. Aerosol Sci. 2015;80:58–74.

[3] Feng Y., Kleinstreuer C., Nicolas C., Rostami A. Computational transport, phase change and deposition analysis of inhaled multicomponent droplet–vapor mixtures in an idealized human upper lung model. J. Aerosol Sci. 2016;96:96–123. DOI: 10.1016/j.jaerosci.2016.03.001.

[4] Kleinstreuer C., Feng Y. Lung deposition analyses of inhaled toxic aerosols in conventional and less harmful cigarette smoke: a review. Int. J. Environ. Res. Public Health 2013;10(9):4454–4485. DOI: 10.3390/ijerph10094454.

[5] Longest P.W., Tian G., Khajeh-Hosseini-Dalasm N., Hindle M. Validating whole-airway CFD predictions of DPI aerosol deposition at multiple flow rates. J. Aerosol Med. Pulm. Drug Deliv. (in press). DOI: 10.1089/jamp.2015.1281.

[6] Ding Y., Kuhlbusch T., Van Tongeren M., Jimenez A.S., Tuinman I., Chen R. Airborne engineered nanomaterials in the workplace—a review of release and worker exposure during nanomaterial production and handling processes. J. Hazard. Mater. (in press). DOI: 10.1016/j.jhazmat.2016.04.075.

[7] Beck-Broichsitter M., Merkel O.M., Kissel T. Controlled pulmonary drug and gene delivery using polymeric nano-carriers. J. Control. Release 2012;161:214–224. DOI: doi: 10.1016/j.jconrel.2011.12.004.

[8] Kleinstreuer C., Feng Y., Childress E. Drug-targeting methodologies with applications: a review. World J. Clin. Cases. 2014;2(12):742–756. DOI: 10.12998/wjcc.v2.i12.742.

[9] Kabilan S., Suffield S.R., Recknagle K.P., Jacob R.E., Einstein D.R., Kuprat A.P. Computational fluid dynamics modeling of Bacillus anthracis spore deposition in rabbit and human respiratory airways. J. Aerosol Sci. 2016; 99:64*#x2013;77. DOI: 10.1016/j.jaerosci.2016.01.011.

[10] Feng Y., Kleinstreuer C. Analysis of non-spherical particle transport in complex internal shear flows. Phys. Fluids. 2013;25:091904. DOI: http://dx.doi.org/10.1063/1.4821812.

[11] Phalen R.F., Raabe O.G. The evolution of inhaled particle dose modeling: a review. J. Aerosol Sci. 2016; 99:7–13. DOI: http://dx.doi.org/10.1016/j.jaerosci.2015.12.008i.

[12] Kleinstreuer C., Zhang Z. Airflow and particle transport in the human respiratory system. Annu. Rev. Fluid Mech. 2010;42:301 334. DOI: 10.1146/annurev-fluid-121108-145453.

[13] Kleinstreuer C., Feng Y. Computational analysis of non-spherical particle transport and deposition in shear flow with application to lung aerosol dynamics—a review. J. Biomech. Eng. 2012;135(2):021008. DOI: 10.1115/1.4023236.

[14] Stuart I.F. Human Physiology. 5th ed. Dubuque, IA, USA: Times Mirror Higher Education Group, Inc.; 1996. 704 p.

[15] Zhang Z., Kleinstreuer C., Kim C.S. Cyclic micron-size particle inhalation and deposition in a triple bifurcation lung airway model. J. Aerosol Sci. 2002;33(2):257–281.

[16] Zhang Z., Kleinstreuer C., Donohue J.F., Kim C.S. Comparison of micro- and nano-size particle depositions in a human upper airway model. J. Aerosol Sci. 2005;36:211–233. DOI: 10.1016/j.jaerosci.2004.08.006.

[17] Li Z., Kleinstreuer C., Zhang Z. Simulation of airflow fields and microparticle deposition in realistic human lung airway models. Part II: particle transport and deposition. Eur. J. Mech. B/Fluids. 2007;26:650–668. DOI: 10.1016/j.euromechflu.2007.02.004.

[18] Zhang Z., Kleinstreuer C. Deposition of naphthalene and tetradecane vapors in models of the human respiratory system. Inhal. Toxicol. 2011;23(1):44–57. DOI: 10.3109/08958378.2010.540261.

[19] Matida E.A., Finlay W.H., Lange C.F., Grgic B. Improved numerical simulation of aerosol deposition in an idealized mouth-throat. J. Aerosol Sci. 2004;35(1):1–19. DOI: 10.1016/S0021-8502(03)00381-1.

[20] Weibel E.R. Morphometry of the human lung. Berlin, Germany: Springer Verlag; 1963. 154 p. DOI: 10.1007/978-3-642-87553-3.

[21] Horsfield K., Dart G., Olson D.E., Filley G.F., Cumming G. Models of the human bronchial tree. J. Appl. Physiol. 1971;31(2):207–217.

[22] Cheng Y., Zhou Y., Chen B.T. Particle deposition in a cast of human oral airways. Aerosol Sci. Tech. 1999;31(4):286–300. DOI: 10.1080/027868299304165.

[23] Heenan A.F., Matida E., Pollard A., Finlay W.H. Experimental measurements and computational modeling of the flow field in an idealized human oropharynx. Exp. Fluids. 2003;35(1):70–84. DOI: 10.1007/s00348-003-0636-7.

[24] Borojeni A.A.T., Noga M.L., Martin A.R., Finlay W.H. An idealized branching airway geometry that mimics average aerosol deposition in pediatric central conducting airways. J. Aerosol Sci. 2015;85:10–16. DOI: dx.doi.org/10.1016/j.jaerosci.2015.03.002.

[25] Sul B., Wallqvist A., Morris M.J., Reifman J., Rakesh V. A computational study of the respiratory airflow characteristics in normal and obstructed human airways. Comput. Biol. Med. 2014;52:130–143. DOI: 10.1016/j.compbiomed.2014.06.008.

[26] Xi J., Longest P.W. Transport and deposition of microaerosols in realistic and simplified models of oral airway. Ann. Biomed. Eng. 2007;35:560–581. DOI: 10.1007/s10439-006-9245-y.

[27] Feng Y. Computational ellipsoidal particle-fluid analysis and discrete element method with applications to particle transport and deposition in human respiratory models [dissertation]. Raleigh, NC, USA: 2013. 585 p.

[28] Stapleton K.W., Guentsch E., Hoskinson M.K., Finlay W.H. On the suitability of k-epsilon turbulence modelling for aerosol deposition in the mouth and throat: a comparison with experiment. J. Aerosol Sci. 2000;31(6):739–749.

[29] Menter F.R. Two-equation eddy-viscosity turbulence models for engineering applications. AIAA J. 1994;32(8):1598–1605.

[30] Zhang Z., Kleinstreuer C. Laminar-to-turbulent fluid–nanoparticle dynamics simulations: model comparisons and nanoparticle-deposition applications. Int. J. Numer. Method. Biomed. Eng. 2011;27(12):1930–1950. DOI: 10.1002/cnm.1447.

[31] Walenga R.L., Longest P.W. Current inhalers deliver very small doses to the lower tracheobronchial airways: assessment of healthy and constricted lungs. J. Pharm. Sci. 2016;105(1):147–159. DOI: 10.1016/j.xphs.2015.11.027.

[32] Tena A.F., Francos J.F., Alvarez E., Casan P. A three dimensional in SILICO model for the simulation of inspiratory and expiratory airflow in humans. Eng. Appl. Comput. Fluid Mech. 2015;9(1):187–198. DOI: 10.1080/19942060.2015.1004819.

[33] Graham D.I., James P.W. Turbulent dispersion of particles using eddy interaction models. Int. J. Multiphase Flow. 1996;22(1):157–175.

[34] Wang L.P., Stock D.E. Stochastic trajectory models for turbulent diffusion: Monte Carlo process versus Markov chains. Atmos. Environ. 1992;26(A):1599–1607.

[35] Kleinstreuer C. Two-phase flow: theory and applications. New York, NY, USA: Taylor & Francis Books, Inc.; 2003. 454 p.

[36] Li M., Ahmadi G. Deposition of aerosols on surfaces in a turbulent channel flow. Int. J. Eng. Sci. 1993;31(3):435–451. DOI: 10.1016/0020-7225(93)90017-O.

[37] Saffman P.G.T. The lift on a small sphere in a slow shear flow. J. Fluid Mech. 1965;22(2): 385–400. DOI: dx.doi.org/10.1017/S0022112065000824.

[38] Feng Y., Kleinstreuer C. Micron-particle transport, interactions and deposition in triple lung-airway bifurcations using a novel modeling approach. J. Aerosol Sci. 2014;71:1–15. DOI: 10.1016/j.jaerosci.2014.01.003.

[39] Turns S.R. An introduction to combustion. New York, NY, USA: McGraw-hill; 2000. 676 p.

[40] Kleinstreuer C., Zhang Z. An adjustable triple-bifurcation unit model for air-particle flow simulations in human tracheobronchial airways. J. Biomech. Eng. 2009;131(2): 021007. DOI: 10.1115/1.3005339.

[41] Tian G., Longest P.W., Su G., Hindle M. Characterization of respiratory drug delivery with enhanced condensational growth using an individual path model of the entire tracheobronchial airways. Ann. Biomed. Eng. 2011;39(3):1136–1153.

[42] Walters D.K., Luke W.H. A method for three-dimensional Navier–Stokes simulations of large-scale regions of the human lung airway. J. Fluids Eng. 2010;132(5):051101. DOI: 10.1115/1.4001448.

[43] Walters D.K., Luke W.H. Computational fluid dynamics simulations of particle deposition in large-scale, multigenerational lung models. J. Biomech. Eng. 2011;133(1): 011003. DOI: 10.1115/1.4002936.

[44] Tena A.F., Clarà P.C. Use of computational fluid dynamics in respiratory medicine. Archivos de Bronconeumología (English Edition). 2015;51(6):293–298. DOI: 10.1016/j.arbr.2015.03.005.

[45] Rowe J.A.R., Burton R., McGregor G., McCauley R., Tang W., Spencer R. Development of a three-dimensional model of the human respiratory system for dosimetric use. Theor. Biol. Med. Modell. 2013;10(1):1. DOI: 10.1186/1742-4682-10-28.

[46] Kitaoka H., Takaki R., Suki B. A three-dimensional model of the human airway tree. J. Appl. Physiol. 1999;87(6):2207–2217.

[47] Burrowes K.S., Swan A.J., Warren N.J., Tawhai M.H. Towards a virtual lung: multi-scale, multi-physics modelling of the pulmonary system. Philos. Trans. R. Soc. Lond. A 2008;366(1879):3247–3263. DOI: 10.1098/rsta.2008.0073.

[48] Lin C.-L., Tawhai M.H., Hoffman E.A. Multiscale image-based modeling and simulation of gas flow and particle transport in the human lungs. Wiley Interdiscip. Rev. Syst. Biol. Med. 2013;5(5):643–655. DOI: 10.1002/wsbm.1234.

[49] Mean flow structures inside the human upper airway. The lattice Boltzmann equation: for fluid dynamics and beyond. Oxford University Press; 2001.

[50] Finck M., Hanel D., Wlokas I. Simulation of nasal flow by lattice Boltzmann methods. Comput. Biol. Med. 2007;37(6):739–749. DOI: 10.1016/j.compbiomed.2006.06.013.

[51] Hörschler I., Schröder W., Meinke M. On the assumption of steadiness of nasal cavity flow. J. Biomech. 2010;43(6):1081–1085. DOI: 10.1016/j.jbiomech.2009.12.008.

[52] Henn T., Thäter G., Dörfler W., Nirschl H., Krause M.J. Parallel dilute particulate flow simulations in the human nasal cavity. Comput. Fluids. 2016;124:197–207.

[53] Krause M.J. Fluid flow simulation and optimisation with lattice Boltzmann methods on high performance computers: application to the human respiratory system [dissertation]. Karlsruhe Institute of Technology: 2010. Available from: http://digbib.ubka.uni-karlsruhe.de/volltexte/1000019768.

[54] Ball C.G., Uddin M., Pollard A. Mean flow structures inside the human upper airway. Flow Turb. Combust. 2008;81(1–2):155–188. DOI: 10.1007/s10494-007-9113-3.

[55] Schroeder W., Eitel G., Soodt Th., Henze A. Analysis of the temporal flow field in a tracheobronchial model. In: J.C.F. Pereira and A. Sequeira, editors. V European Conference on Computational Fluid Dynamics; 14–17 June 2010; Lisbon, Portugal. 2010. p. 1–13.

[56] Eite G., Freitas R.K., Lintermann A., Meinke M., Schröder W. Numerical simulation of nasal cavity flow based on a Lattice-Boltzmann method. In: Dillmann A., Heller G., Klaas M., Kreplin H.-P., Nitsche W., Schroeder W., editors. New Results in Numerical and Experimental Fluid Mechanics VII. Springer Berlin Heidelberg; 2010. p. 513–520. DOI: 10.1007/978-3-642-14243-7_63.

[57] Imai Y., Miki T., Ishikawa T., Aoki T., Yamaguchi T. Deposition of micrometer particles in pulmonary airways during inhalation and breath holding. J. Biomech. 2012;45(10): 1809–1815. DOI: 10.1016/j.jbiomech.2012.04.017.

[58] Wang Y., Elghobashi S. On locating the obstruction in the upper airway via numerical simulation. Respir. Physiol. Neurobiol. 2014;193:1–10. DOI: 10.1016/j.resp.2013.12.009.

[59] Chen J., Gutmark E. Numerical investigation of airflow in an idealized human extra-thoracic airway: a comparison study. Biomech. Model. Mechanobiol. 2014;13(1):205–214. DOI: 10.1007/s10237-013-0496-x.

[60] Trunk R., Henn T., Dörfler W., Nirschl H., Krause M.J. Inertial dilute particulate fluid flow simulations with an Euler–Euler lattice Boltzmann method. J. Comput. Sci. (in press). DOI: 10.1016/j.jocs.2016.03.013.

[61] Li Z., Kleinstreuer C. Airflow analysis in the alveolar region using the lattice-Boltzmann method. Med. Biol. Eng. Comput. 2011;49(4):411–451. DOI: 10.1007/s11517-011-0743-1.

[62] Aidun C.K., Jonathan R.C. Lattice-Boltzmann method for complex flows. Annu. Rev. Fluid Mech. 2010;42:439–472. DOI: 10.1146/annurev-fluid-121108-145519.

[63] Ladd A.J. Numerical simulations of particulate suspensions via a discretized Boltzmann equation. Part 1. Theoretical foundation. J. Fluid Mech. 1994;271(1):285–309. DOI: http://dx.doi.org/10.1017/S0022112094001771.

[64] Ladd A.J. Numerical simulations of particulate suspensions via a discretized Boltzmann equation. Part 2. Numerical results. J. Fluid Mech. 1994;271:311–339. DOI: http://dx.doi.org/10.1017/S0022112094001783.

[65] Mei R., Yu D., Shyy W., Luo L.S. Force evaluation in the lattice Boltzmann method involving curved geometry. Phys. Rev. E. 2002;65(4):041203. DOI: http://dx.doi.org/10.1103/PhysRevE.65.041203.

[66] Khalil F., Läer S. Physiologically based pharmacokinetic modeling: methodology, applications, and limitations with a focus on its role in pediatric drug development. J. Biomed. Biotechnol. 2011;2011:907461. DOI: 10.1155/2011/907461.

[67] Clewell H.J., Reddy M.B., Lave T., Andersen M.E. Physiologically Based Pharmacokinetic Modeling. In: Pharmaceutical Sciences Encyclopedia. John Wiley & Sons, Inc.; 2010. p. 1–62. DOI: 10.1002/9780470571224.pse065.

[68] Krishnan K., Peyret T. Physiologically based toxicokinetic (PBTK) modeling in ecotoxicology. In: Ecotoxicology Modeling. Springer, US; 2009. p. 145–175.

[69] Rigg A., Hindle M., Longest P.W. Absorption and clearance of pharmaceutical aerosol in the human nose: effects of nasal spray suspension particle size and properties. Pharm. Res. 2016;33:909–921. DOI: 10.1007/s11095-015-1837-5.

[70] Corley R., Kabilan S., Kuprat A., Carson J., Minard K., Jacob R., Timchalk C., Glenny R., Pipavath S., Cox T., Wallis C., Larson R., Fanucchi M., Postlethwait E., Einstein D. Comparative computational modeling of airflows and vapor dosimetry in the respiratory tracts of a rat, monkey, and human. Toxicol. Sci. 2012;ksf168. DOI: 10.1093/toxsci/kfs168.

[71] Schroeter J.D., Kimbell J.S., Gross E.A., Willson G.A., Dorman D.C., Tan Y.M., Clewell H.J. Application of physiological computational fluid dynamics models to predict interspecies nasal dosimetry of inhaled acrolein. Inhal. Toxicol. 2008;20:227–243. DOI: 10.1080/08958370701864235.

[72] Corley R.A., Kabilan S., Kuprat A.P., Carson J.P., Jacob R.E., Minard K.R., Einstein D.R. Comparative risks of aldehyde constituents in cigarette smoke using transient computational fluid dynamics/physiologically based pharmacokinetic models of the rate and human respiratory tracts. Toxicol. Sci. 2015;kfv072. DOI: 10.1093/toxsci/kfv071.

[73] Boger E., Evans N., Chappell M., Lundqvist A., Ewing P., Wigenborg A., Friden M. Systems pharmacology approach for prediction of pulmonary and systemic pharma-

cokinetics and receptor occupancy of inhaled drugs. CPT Pharmacometr. Syst. Pharmacol. 2016;5:201–210. DOI: 10.1002/psp4.12074.

[74] Weber B., Hochhaus G. A pharmacokinetic simulation tool for inhaled corticosteroids. AAPS J. 2013;15(1):159–171. DOI: 10.1208/s12248-012-9420-z.

[75] Kolanjiyil A.V., Kleinstreuer C. Nanoparticle mass transfer from lung airways to systemic regions – Part II: multi-compartmental modeling. J. Biomech. Eng. 2013;135(12):121004.

[76] Caldwell J.C., Evans M.V., Krishnan K. Cutting edge PBPK models and analyses: providing the basis for future modeling efforts and bridges to emerging toxicology paradigms. J. Toxicol. 2012;2012:852384. DOI: 10.1155/2012/852384.

[77] Jamei M. Recent advances in development and application of physiologically-based pharmacokinetic (PBPK) models: a transition from academic curiosity to regulatory acceptance. Curr. Pharmacol. Rep. 2016;2:161–169. DOI: 10.1007/s40495-016-0059-9.

[78] Crowe C.T., Troutt T.R., Chung J.N. Numerical models for two-phase turbulent flows. Annu. Rev. Fluid Mech. 1996;28:11–43.

[79] Crowe C.T., Schwarzkopf J.D., Sommerfeld M., Tsuji Y. Multiphase Flows with Droplets and Particles. 2nd ed. Boca Raton, FL, USA: CRC Press; 2011.

[80] Zhong W., Yu A., Liu X., Tong Z., Zhang H. DEM/CFD-DEM modelling of non-spherical particulate systems: theoretical developments and applications. Powder Technol. 2016; 302:108–152. DOI: 10.1016/j.powtec.2016.07.010.

[81] Cundall P.A., Strack O.D.L. A discrete numerical model for granular assemblies. Geotechnique 1979;29:47–65.

[82] Wang G.Q., Hao W.J., Wang J.X. Discrete element method and its application using EDEM. China: Northwestern Polytechnical University Press; 2010.

[83] O'Sullivan C. Particulate discrete element modelling: a geomechanics perspective. CRC Press; 2011.

[84] Ren B., Zhong W., Chen Y., Chen X., Jin B., Yuan Z., Lu Y. CFD-DEM simulation of spouting of corn-shaped particles. Particuology. 2012;10(5):562–572. DOI: 10.1016/j.partic.2012.03.011.

Chapter 4

Lidar Mapping of Near-Surface Aerosol Fields

Tanja Dreischuh, Ivan Grigorov, Zahary Peshev,
Atanaska Deleva, Georgi Kolarov and
Dimitar Stoyanov

Additional information is available at the end of the chapter

http://dx.doi.org/10.5772/65274

Abstract

Near-surface atmospheric measurements over urban or industrial areas aimed at assisting the air-quality monitoring attain increasing societal significance due to the strong and direct impact of aerosol pollutions in the low troposphere on the human health. In this chapter, we present experimental results on lidar mapping of aerosol fields over the city of Sofia (Bulgaria), its suburbs and adjacent towns and villages, obtained during an extensive 7-month experimental campaign in 2015. The measurements are conducted by scanning observation zones in horizontal and vertical directions using lidar systems developed at the Institute of Electronics, Bulgarian Academy of Sciences. Based on the aerosol backscattering profiles retrieved at different azimuth or elevation angles, two-dimensional color-coded sector maps of the near-surface aerosol density are obtained, overlaid on the topological map of the Sofia region. The analysis of the lidar maps shows good correlation between the aerosol density distribution and the locations of important sources of aerosol pollutions in the zones of observation, such as city streets with intense traffic, industrial facilities, densely populated residential districts, etc. The results reported demonstrate that aerosol lidar mapping could be regarded as an effective approach for accurate and reliable determination of the density, spatial distribution, and temporal dynamics of close-to-ground aerosols, covering broad urban areas. Possibilities of incorporating synergistically lidar mapping technologies into municipal air-quality monitoring systems are also discussed.

Keywords: aerosol, lidar, scanning lidar, near-surface aerosol lidar mapping, air-quality monitoring

1. Introduction

The poor ecological state of the environment is a serious global problem which arose as a result of the rapid development of industry, agriculture, urbanization, and transport. Significant factors affecting the state of the environment are both natural (hydrometeors, volcanic ash, desert dust, smoke aerosols, etc.) and anthropogenic (industrial, fuel combustion, fires, heating, etc.) atmospheric aerosols or particulate matter (PM) [1]. The aerosols are fine solid or liquid particles, determining largely the climate, temperature, and dynamic structure of the atmosphere, the functioning of ecosystems, the microphysical properties of clouds and various chemical and photochemical processes in air [2]. The concentration of aerosols in the atmosphere determines the air quality, and in turn affects the human health [[3] and references therein]. The fine and ultrafine aerosol particles are particularly harmful for human health, as they more easily penetrate and accumulate in the human body and lead to an increase in cardiovascular and respiratory diseases and even to lung cancer [4, 5]. Because of the above facts and considerations, atmospheric aerosols have been the subject of more intensive research in recent decades [6, 7]. Particularly, near-surface atmospheric measurements over densely populated or industrial areas, purposed to help monitor air quality, have attained increasing societal significance.

Laser radars (lidars) are recognized to be a reliable and powerful instrument for investigating atmospheric objects and air parameters [7–9]. As compared to other measurement approaches, the lidar technique exhibits advantages such as possibilities for performing fast, highly sensitive and accurate monitoring of vast atmospheric domains with high spatial and temporal resolution. Lidar systems are mainly used to assess the vertical structure of the aerosol layers and determine the optical and microphysical properties of the vertical profiles of the aerosols. Scanning ground-based and airborne lidars are applied to produce three-dimensional (3D) maps of earth's surface and man-made features [10], as well as for characterization of tropospheric wind profiles [11, 12] and temperature fields [13]. Along with the above applications, scanning elastic-scattering lidars are used to obtain maps of important atmospheric pollutants, particularly, of near-surface aerosol fields [14–17]. The aerosol lidar mapping represents a fast and effective approach to detect polluting aerosol loads over broad areas, as well as to characterize them in terms of local density, spatial distribution, and temporal dynamics [18–20].

In this chapter, we present experimental results on lidar mapping of near-surface atmospheric aerosol fields over the city of Sofia, its suburbs and surrounding villages, obtained during an extensive 7-month experimental campaign in 2015 [19]. It was carried out in the framework of a common project with Sofia Municipality aimed to help the local authorities to improve the regional air-quality monitoring. Possibilities are also discussed to incorporate lidar mapping technologies synergistically into municipal air-quality monitoring systems. Aerosol lidar maps are considered to become basic components of such monitoring systems. Their advantages result from the high efficiency of the laser light interaction with the atmospheric particles, thus, providing better visualization of atmospheric motions in comparison with other remote sensing techniques, such as the microwave or acoustic probing.

The analysis of the lidar mapping experiments performed was focused on the following issues:

i. Two-dimensional (2D) aerosol density distribution above the city areas.

ii. Temporal and spatial dynamics of the near-surface aerosol fields.

iii. Estimating effects of the city structure and terrain topography on the aerosol distribution obtained.

The chapter is organized as follows: In Section 2, we describe the experimental instrumentation, data processing/visualization approaches, and the area investigated. Then, in Section 3, series of experimental data on lidar aerosol mapping over the Sofia region are presented. Results are reported and analyzed of horizontal and vertical lidar scanning of the aerosol density distribution, as well as its temporal dynamics. Special attention is paid in Section 4 to the synergistic effects of mutual use of lidar aerosol mapping and in-situ measurement sites of the air pollution. The main results and conclusions are summarized in Section 5.

2. Experimental setup

2.1. Description of the area investigated

Sofia (the capital of Bulgaria) is located at about 550 m above sea level (a.s.l.), in a valley surrounded by hills and mountains, bordering Stara Planina to the northeast and Vitosha to the southwest. This topographic position, and the temperature inversion conditioned by it, is one of the factors determining the regional air quality. In the past years, the air pollution over the city of Sofia has become a serious ecological problem, provoked by the presence of different industrial facilities, a considerable decrease of green zones, as well as the accelerating growth of the population and the number of cars. The analysis of the PM amount and properties over Europe [21] showed that the aerosol concentrations have reached second-highest levels in eastern and southern Europe.

The measurement site (42.65N, 23.38E; 590 m a.s.l.) is located at the Institute of Electronics, Bulgarian Academy of Sciences, in the southeast part of Sofia. **Figure 1** shows a topographical map of Sofia area, overlaid by lidar sector scans in north-northwest (north-NW) (to the central parts of Sofia) and southwest (SW) (toward Vitosha Mountain) directions. In the former case, a horizontal lidar scanning (at a step of 1.7°) was performed in two close azimuth sectors of 8.5°(321–329.5° with respect to the north clockwise) and 17° (348–5°). In the latter case, the investigated sector was 50° (188–238°, at a step of 2°), whereas a low elevation angle (6–7°) was used to measure near-surface atmospheric aerosol fields because of the presence of high buildings in the angular sectors of measurements.

Under the conditions mentioned above, surface areas of about 100 km^2 were scanned and mapped over the central city zone, the north industrial zone and the south urban and suburb parts, including the north slopes of Vitosha Mountain. The results of the lidar aerosol mapping over these areas would allow one to detect and analyze the aerosols of different origin (natural, urban, industrial, etc.), as well as to contribute to the establishment of a modern city air-quality monitoring system.

Figure 1. Topographical map of Sofia region with outlined azimuth sectors of lidar scanning.

2.2. Lidar systems

The measurements described in this chapter were conducted by scanning observation zones in horizontal and vertical directions over Sofia using lidar systems developed at the Laser Radars Laboratory of the Institute of Electronics, Bulgarian Academy of Sciences (LRL-IE). The LRL-IE working groups actively participate in the joint lidar research over the European continent in the framework of the European Aerosol Research Lidar Network, performing systematic lidar monitoring of atmospheric processes [18, 22], unusually high concentrations of aerosols in the troposphere [23], transport of mineral dust from Sahara desert [24], volcanic eruptions [25], and formation of smoke layers resulting from forest or industrial fires [26].

The lidar systems used in the experimental campaign considered are shown in **Figure 2**. The first one is based on a Cu-vapor laser emitting pulses with duration of 10 ns at a repetition rate of 5 kHz and wavelengths of 510.6 and 578.2 nm. Their mean powers are 1.2 and 0.8 W, respectively. The beam divergence is about 2 mrad. The laser beam is directed in parallel to

(a) (b)

Figure 2. Photographs of the Cu-vapor (a) and Nd:YAG (b) laser-based lidars used in the aerosol lidar mapping experiments.

the axis of the receiving telescope, forming a lidar base of ~30 cm between the axes. A Cassegrain-type telescope with 20 cm aperture and 1 m focal length receives the backscattered laser emission from the atmosphere. Narrow-band interference filters are used to separate the lidar signals. Registration in a photon-counting mode is applied. The single electron pulses, produced by the photodetector (a photomultiplier), are accumulated by a photon-counting board in a computer. This board allows registration of the backscattered lidar signal with a spatial resolution of 15 or 30 m in 1024 samples and variable averaging time. In the present experiments, the lidar profiles of the laser emission backscattered in the atmosphere were registered with an accumulation time of 1 min. In addition, averaging was performed by summation of the data of 5–10 profiles; thus, the effective measurement time for each profile amounted to 5–10 min in a single azimuth direction.

The lidar used to perform lidar mapping experiments in the south-southwest direction is based on a solid-state frequency-doubled Nd:YAG laser (pulse energy of up to 600 mJ at 1064 nm, 80 mJ at 532 nm; fixed repetition rate of 2 Hz, FWHM pulse duration of 15 ns, beam divergence of 2 mrad), acting as a two-wavelength lidar transmitter. The optical part of the lidar receiver consists of a Cassegrain-type telescope (aperture 35 cm; focal distance 200 cm) and a three-channel spectrum analyzer based on narrow-band interference filters (1–3 nm FWHM). The receiver's electronic part comprises three compact photoelectronic modules, each including a photodetector, a 10 MHz 14-bit analog-to-digital converter (ADC), a high-voltage power supply, and controlling electronics. The signals backscattered in the atmosphere are digitized every 100 ns by the ADC, resulting in a 15 m range resolution. The system provides detection and storage of lidar returns from distances of up to 30 km. The lidar is mounted on a stable metal coaxial construction allowing reliable fixing and precise synchronized mutual motion of both the telescope and the output laser beam in horizontal and vertical direction with an angular resolution of about 1°.

2.3. Lidar data processing: deriving the atmospheric aerosol backscatter profiles

The lidar remote sensing is based on the interaction (absorption and scattering) of the laser light with molecules and aerosols in the atmosphere. The detected backscattered lidar signals contain information concerning the state and composition of the probed atmospheric domain. The so-called lidar equation describes the power of the received backscattered signal as a range-resolved function of the lidar parameters and the atmospheric optical properties (aerosol backscattering and extinction coefficients). For a single-scattering elastic lidar (measuring backscattered light at the same wavelength as the sensing laser wavelength λ) the power $P(r)$, detected at a time t after the instant of pulse emission, is written as [27]:

$$P(r) = P_0 \frac{c\tau}{2} A\varepsilon \frac{\gamma(r)}{r^2} [\beta_a(r) + \beta_m(r)] \exp\{-2\int_0^r [\alpha_a(\rho) + \alpha_m(\rho)] d\rho\}, \tag{1}$$

where P_0 is the average power of a single laser pulse, c is the speed of light, $r = ct/2$ is the distance along the laser beam path, τ is the pulse duration, A is the area of the receiver, ε is the overall system efficiency, $\gamma(r)$ describes the overlap between the laser beam and the receiver field of view, and $\beta_a(r)$, $\beta_m(r)$ and $\alpha_a(r)$, $\alpha_m(r)$ are the backscattering and extinction coefficients for aerosols and molecules, respectively, at wavelength λ.

The determination of the aerosol extinction and backscattering coefficients (BSCs) on the basis of Eq. (1) requires the solution of a Bernoulli differential equation. A stable solution has been proposed by Klett [28] and Fernald [29], applying an inverse integration algorithm starting from the far end of the lidar sounding path. In the case of the backscattering coefficient (BSC), it has the following form:

$$\beta_a(r) = -\beta_m(r) + \frac{P(r) r^2 \exp\{-2[S_a(r) - S_m] \int_r^{r_{ref}} \beta_m(\rho) d\rho\}}{\frac{P(r_{ref}) r_{ref}^2}{\beta_a(r_{ref}) + \beta_m(r_{ref})} + 2\int_r^{r_{ref}} S_a(\rho) P(\rho) \rho^2 \exp\{-2[S_a(\rho) - S_m] \int_\rho^{r_{ref}} \beta_m(\rho') d\rho'\} d\rho} \tag{2}$$

where $S_a(r) = \alpha_a(r)/\beta_a(r)$ and $S_m = \alpha_m(r)/\beta_m(r) = 8\pi/3$ are the aerosol and the molecular extinction-to-backscatter lidar ratios, respectively. The reference range r_{ref} is chosen so that the aerosol backscatter coefficient at that point is either negligible compared to the molecular backscatter coefficient or is known from other sources. The vertical profiles of the $\beta_m(r)$ could be determined from the Standard Atmosphere Model [30] and from meteorological data. This algorithm is now widely applied in practice, assuming also that the aerosol lidar ratio S_a is invariant along the laser beam path. The exact value of this ratio is determined depending on the laser wavelength and also on *a priori* assumptions about the atmospheric conditions and the type of the aerosols observed.

In the case of lidar measurements in vertical or quasi-vertical directions, aerosol-free atmospheric domains are usually reached at certain altitudes (as a rule higher than 5–6 km) in the free troposphere, where the total backscatter coefficient $\beta(r) = \beta_a(r) + \beta_m(r) \approx \beta_m(r)$ is *a priori* known. In the case considered here of horizontal or quasi-horizontal lidar measurements, lidar paths pass through close-to-surface atmospheric parts rich in aerosols, for which molecular reference values of $\beta(r)$ could not be used. In such cases, alternative and/or auxiliary approaches for retrieving the profiles of $\beta(r)$ have to be applied, in order to characterize the aerosol content in the observation areas. Such an approach is the so-called "slope method" [27], applicable to characterizing atmospheric domains with relatively homogeneous aerosol composition and concentration.

A typical feature of the atmosphere is its vertical stratification, expressed in the formation of a vertical succession of horizontally extended layers of different thickness. Inside these layers, the atmospheric air content and parameters remain practically constant over considerable horizontal distances. The longer the reachable lidar range of horizontal sounding, the higher the probability such homogeneous air volumes to be present along the lidar line of sight, providing favorable conditions for the method to be used. Thus, the slope method appears to be very suitable for determining aerosol characteristics in horizontal lidar measurements or such performed at low-elevation angles. The accuracy of this method increases with increasing the aerosol concentration, favoring its application to the lidar measurements conducted in the near-surface atmospheric layers where the highest aerosol concentrations are usually observed.

Applying the slope method to solving the lidar equation, one can obtain the following expressions for the aerosol extinction and backscattering coefficients:

$$\alpha_a(r) = \alpha_a = -0.5d\{\ln[P(r)r^2]\}/dr \tag{3}$$

and

$$\beta_a(r) = \beta_a = \alpha_a/S_a \tag{4}$$

Within the mentioned atmospheric parts of homogeneous aerosol parameters, S_a keeps empirically defined constant values depending on the aerosol types and densities. Both accuracy and reliability of the slope method increase proportionally to the lengths of the homogeneous parts present along the lidar beam path.

As an important advantage of the slope method, in comparison to other lidar approaches, determination of the aerosol extinction or backscattering is only or predominantly based on lidar measurement data, without the need of information or suppositions concerning relations between the analyzed quantities. In addition, the method makes use of simple mathematics, provides analytical solutions, and does not require numerical approaches and algorithms.

To implement the lidar mapping described here of the near-surface aerosol density distribution over Sofia region, a combination of the widely adopted and well-elaborated method of Klett-Fernald and the slope method was used. In this combination, the slope method is applied to determining the aerosol extinction and backscattering coefficients in appropriate parts of the lidar beam path by using the technology presented above (Eqs. (3) and (4)). Subsequently, the values of $\beta_a(r)$ obtained are used as reference (calibrating) ones in Eq. (2), in retrieving the whole range profiles of $\beta_a(r)$ by means of the Klett-Fernald approach. In this manner, the advantages of the two approaches are synergistically combined. As a result, the lidar range profiles of the aerosol extinction and backscatter coefficients are retrieved with relatively high

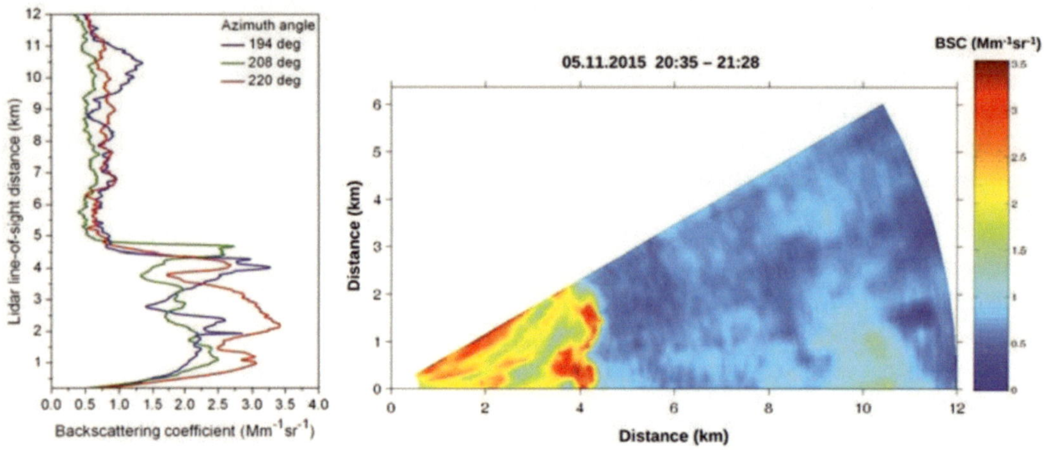

(a) (b)

Figure 3. Range profiles of the aerosol backscattering coefficient at three different azimuth angles (a) and aerosol distribution lidar map based on a series of BCS profiles (b) as measured in the time interval 20:35–21:28 LT on 5 November 2015.

precision and reliability, which are accordingly transferred to the colormaps based on them of the near-surface aerosol density distribution.

2.4. Lidar mapping of aerosol fields

Generally, the aerosol field could be described as a distribution of the aerosol mass concentration M (μg/m^3) defined as the mass of PM per unit volume. From the lidar measurements, the extinction and backscattering coefficients of the aerosol particles are determined that are directly proportional to the aerosol mass concentration:

$$M = k\alpha_a = k\beta_a S_a. \tag{5}$$

The mass concentration could be retrieved from the lidar data combining different experimental and numerical approaches [31]. So, obtaining data about the distribution of the aerosol backscattering coefficient could be regarded as representative for the aerosol mass concentration distribution.

Figure 3 shows an example of the stages of formation of an aerosol lidar map using measurements performed on 5 November 2015, in the time interval 20:35–21:28 local time (LT). Aerosol backscattering profiles obtained at different azimuth angles along a fixed elevation angle are presented in **Figure 3(a)**. On the basis of a series of such profiles, 2D color-coded sector maps of the near-surface aerosol density could be created. In **Figure 3(b)**, an aerosol lidar map is displayed in Cartesian coordinates, based on the entire set of BSC profiles in the azimuth sector 190–220°, including the ones in **Figure 3(a)**. Finally, the sector maps so-obtained are superposed on the satellite maps of the corresponding city region.

http://dx.doi.org/10.5772/65274

3. Experimental results and discussion

Results of lidar measurements and mapping of the near-surface atmospheric aerosol fields over the city of Sofia, suburbs, and surrounding villages, obtained during the experimental campaign in 2015, are shown and discussed below. Lidar maps are presented from vertical and horizontal scanning of the areas investigated.

3.1. Mapping of aerosol fields by vertical scanning

In order to acquire detailed information about the vertical structure of the aerosol concentration, lidars perform vertical slice scans. The lidar data shown on **Figure 4** represent a two-dimensional color-coded sector map of the aerosol density distribution within the scanned volume of the atmosphere. The map is constructed using lidar profiles (averaged over five individual scans) obtained along a fixed azimuth in NW-direction at different elevation angles (0–10°), with an increment of 1°. The horizontal direction of the lidar scanning, covering

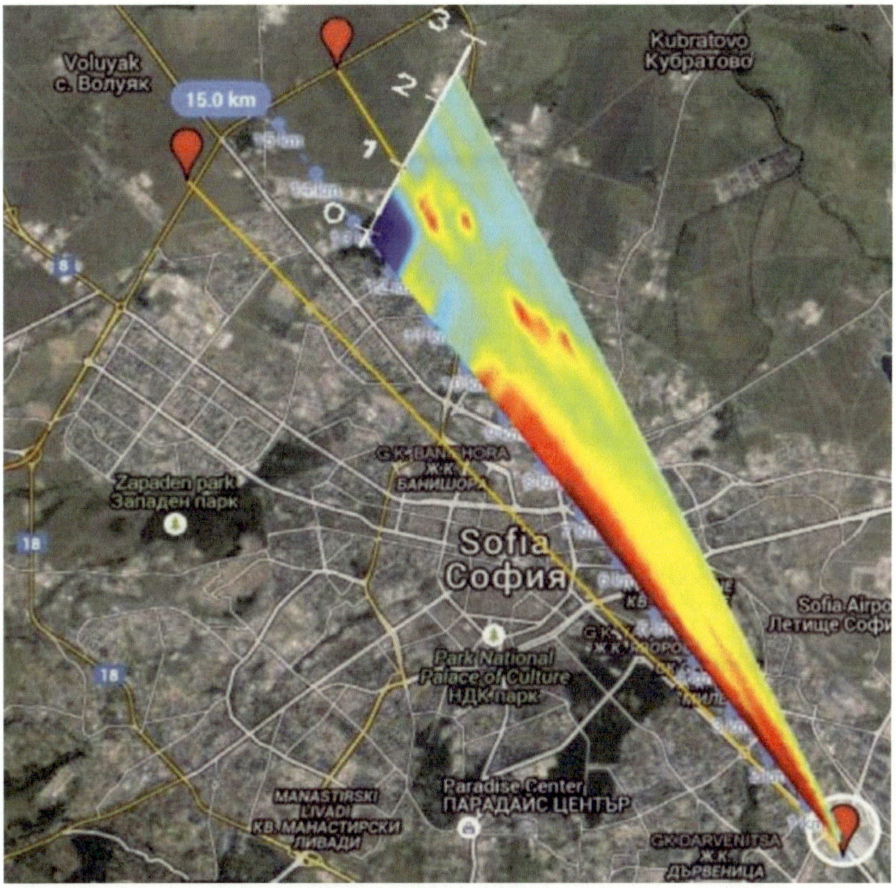

Figure 4. Color-coded sector map of the vertical aerosol density distribution obtained along a fixed azimuth in NW-direction (326° with respect to the north clockwise).

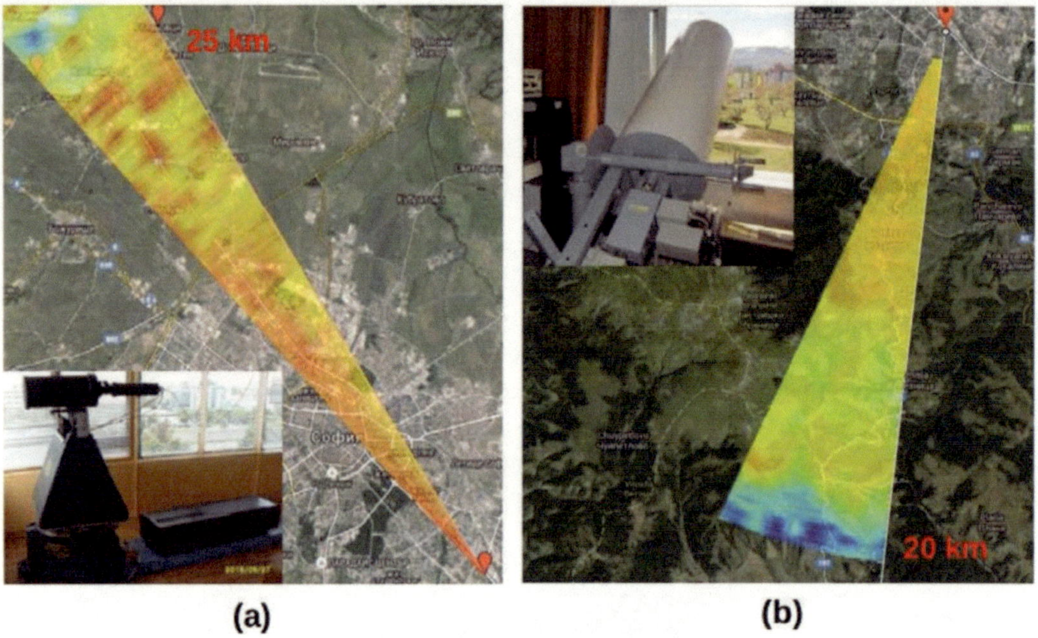

Figure 5. Colormaps of the near-surface aerosol density distribution demonstrating the maximum achievable operational distances in NW (a) and SW (b) directions, performed by the lidar systems shown in the insets.

distances of about 12 km was close to a thoroughfare with intense traffic. The vertical structure of the aerosol density of the atmosphere is clearly visible on the map. A well-pronounced vertical layer near ground surface was observed at a height in the range of 500–700 m, located above the city away from the lidar station until beyond the city center. At a height of ~1 km above ground, aerosol formations were observed with a density exceeding that of the ambient atmosphere, probably low clouds. Thus, this vertical map demonstrates the capability of such a type of lidar measurements to determine quickly and efficiently the location of the sources of anthropogenic PM emission in the atmosphere. On the other hand, it is clear that the horizontal scanning lidar measurements, made at a low altitude in the range of 500–700 m, provide sufficient information about the air pollution and near-ground surface aerosol fields.

3.2. Range limits of lidar measurements

Figure 5 illustrates the range limits of lidar measurements in NW (**Figure 5(a)**) and SW (**Figure 5(b)**) directions, performed by lidars with a Cu-vapor laser and a Nd:YAG laser, respectively. In the first case, the operational distance was from 900 m to 25–28 km in nighttime and decreased to about 10–15 km in daytime, due to intensive sky illumination. The maximum distance was limited by the high laser pulse repetition rate, because of an overlap of the laser pulse scattered from far away with the next pulse scattered from a close distance. In SW-direction to the Vitosha Mountain, the assessed maximum distance was longer than 20 km as determined by the surface topology in the observation area.

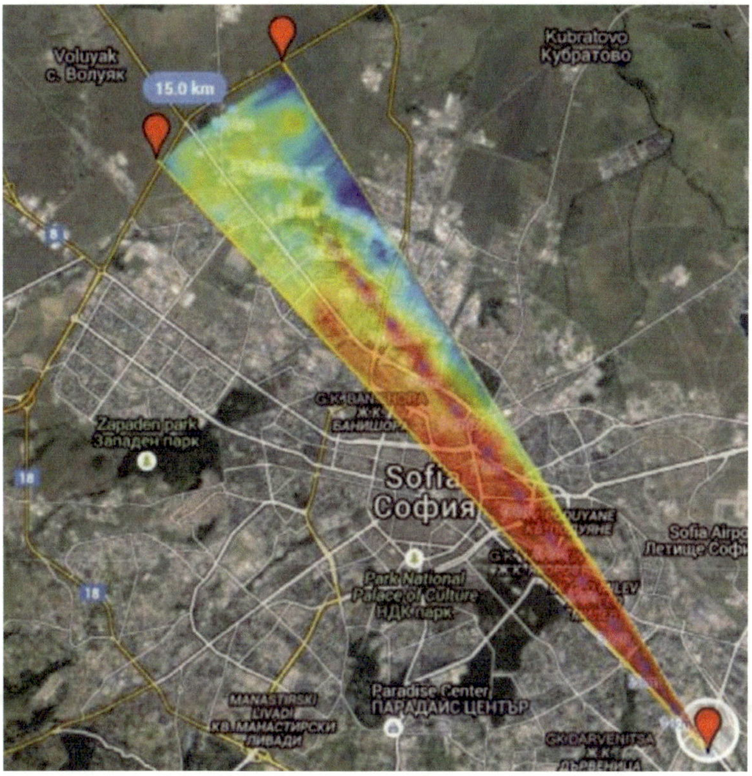

Figure 6. Colormap of the aerosol density distribution as measured on 27 July 2015 in the time interval 21:33–22:10 LT, at distances of up to 15 km.

3.3. Lidar mapping of aerosol fields over Sofia's central parts

Lidar monitoring and mapping of the near-surface aerosols in the atmosphere above the central parts of Sofia were performed by the lidar equipped with a Cu-vapor laser at the wavelength of 510.6 nm.

Figure 6 presents the results of lidar measurements carried out on 27 July 2015, at 21:33–22:10 LT, when relatively strong air pollution was observed. The distance covered by the lidar sounding was 15 km in an azimuth sector of 8.5°. A dust cloud was observed in the atmosphere near ground surface over most of the observation zone. Only the blue-colored areas, at 1.5–2 km away from the two large boulevards, showed a lower concentration of dust particles in the air. The specific movement of air masses, from SW-to-NE-direction, causes a mixing of dust pollutants into the larger part of the area over the city observed by the lidar. The values measured of the aerosol BSC are in the order of $0.5–8 \times 10^{-6}$ $m^{-1}sr^{-1}$.

Figure 7 presents results of lidar mapping in the two main sectors of scanning over the central parts of Sofia, performed in different time periods during the measurement campaign and exhibiting similar features of the aerosol distribution, in particular, the influence of populated areas on the aerosol density. The near-surface aerosol distribution, resulting from lidar

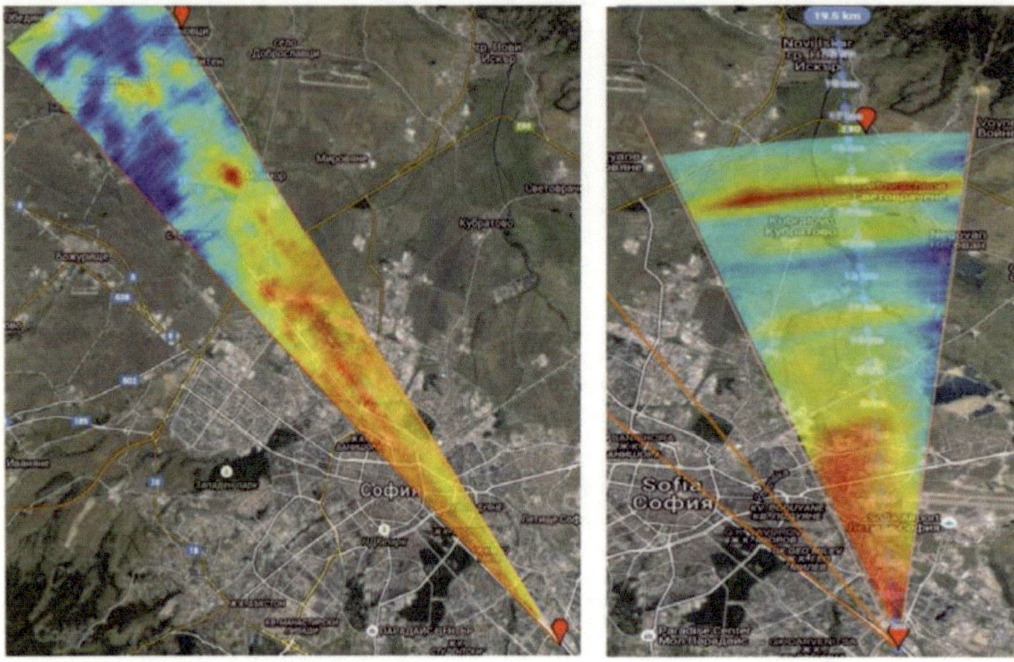

(a) (b)

Figure 7. Color-coded maps of the near-surface aerosol density distribution as measured on 5 August 2015, at 22:10–22:50 LT (a) and on 7 October 2015 in the time interval 19:35–20:30 LT (b).

measurements performed on 5 August 2015, at 22:10–22:50 LT, is shown in **Figure 7(a)**, covering a distance of 25 km. Due to the heavy city traffic, relatively higher values of the aerosol BSC were observed close to the busy streets and over the entire central parts reaching the city ring road. At distances beyond the ring road, the aerosol pollution concentration dropped rapidly, except for some areas near two local villages. We, therefore, assumed that the aerosol fields observed by the lidar were of anthropogenic origin. The values measured of the aerosol BSC were in order of $0.3–4.3 \times 10^{-6}$ $m^{-1}sr^{-1}$. **Figure 7(b)** presents a map of a lidar scanning conducted on 7 October 2015 within the NW sector, in the time interval 19:35–20:30 LT. The sounding comprised 11 successive scans in an angular sector of 17° by an angle step of 1.7°, covering a distance of 16 km. Well-defined areas of higher aerosol pollution were visible over the city areas, as well as over some residential districts in the far measurement zone. The values calculated of the atmospheric BSC were from 0.5 to 9.8×10^{-6} $m^{-1}sr^{-1}$.

In **Figure 8**, results are presented of lidar measurements in the two main sectors of scanning as in **Figure 7**, performed consecutively in the time intervals 19:22–19:48 and 19:54–20:54 LT on 4 November 2015. The first measurement (shown at the left angular sector of **Figure 8**) directed northwestward covered the central city zones including and being nearly parallel to one of the main city thoroughfare, reaching distances of up to 12 km. The second lidar sounding (shown at the right angular sector of **Figure 8**) was directed north-northwestward to distances of 13 km, covering densely populated residential districts in the part of the map near the lidar.

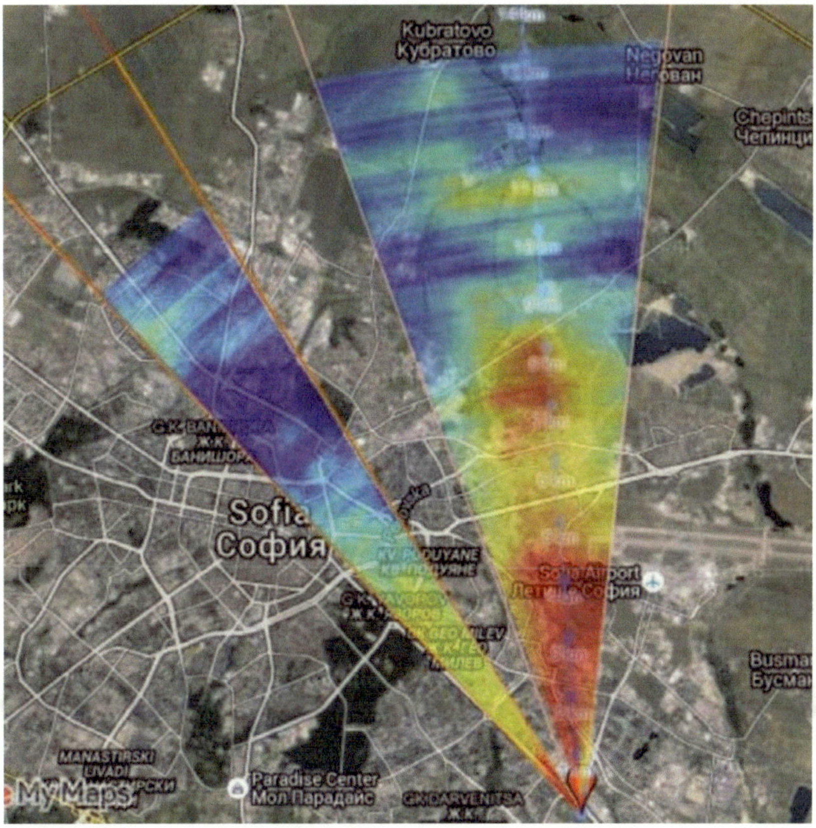

Figure 8. Color-coded maps of the near-surface aerosol density distribution as measured on 4 November 2015, in the time intervals 19:22–19:48 LT (left angular sector) and 19:54–20:54 LT (right angular sector).

Along this second direction, an area of high aerosol concentration was observed, which extended to a distance of 5 km with respect to the lidar. Another area of high concentration of the near-surface aerosols was observed at a distance of 7–8 km, at the end of the urban area. At greater distances, the aerosol air pollution observed was negligible, as shown by the green-blue colors in the figure. The BSC values calculated ranged from 0.5 to $6.2 \times 10^{-6}\ m^{-1}sr^{-1}$.

In order to demonstrate the capability of the lidar aerosol mapping technology applied to follow the temporal evolution of the near-surface aerosol density distribution, a series of successive lidar scans over the same areas were carried out. **Figure 9** presents four lidar maps resulting from measurements conducted on 18 November 2015, as averaged over 30 min intervals. The start times of each measurement are marked in the upper right corner of the corresponding figure panels. The measurements were implemented in the NW angular sector, reaching distances of up to 9.5 km. These maps illustrate the changes occurring in the near-surface aerosol fields measured over the city area in the observation zones. As can be seen, the areas located near the main city thoroughfare with the most intense traffic are colored in red-brown, indicating strong aerosol pollution, probably due to the car exhaust emissions. Inspecting the four pictures

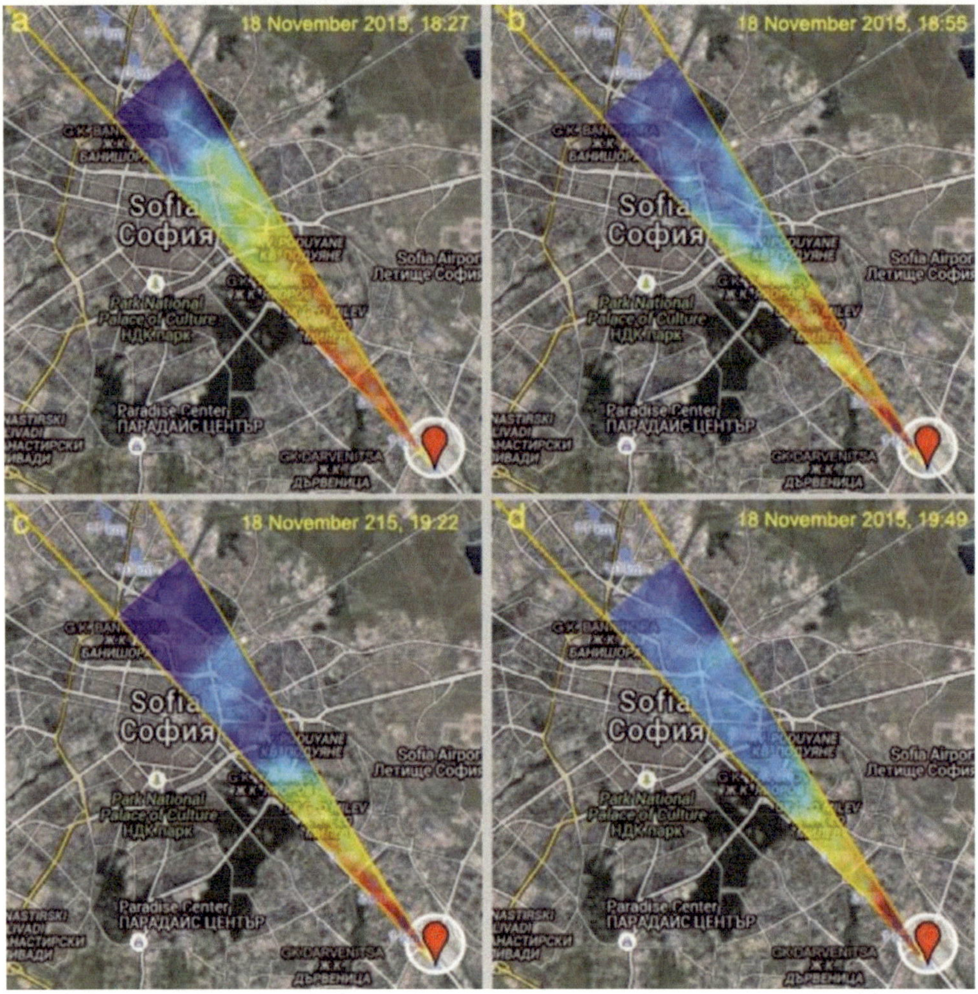

Figure 9. Color-coded maps of the aerosol density distribution as measured on 18 November 2015 in the time intervals 18:2718:52 LT (a), 18:55–19:20 LT (b), 19:22–19:47 LT (c), and 19:49–19:14 LT (d).

presented in their chronological order, one can perceive a progressive shrinking of the part of the maps polluted by aerosols. This peculiarity can be ascribed to the progressively diminishing traffic intensity in the evening hours, resulting in less car aerosol emissions.

3.4. Lidar mapping of aerosol fields toward Vitosha Mountain

Series of both daytime and nighttime lidar measurements of the near-surface aerosol density distribution were carried out in the period 3–9 November 2015. The meteorological conditions during the measurements were as follows: a relatively high temperature for the season (18–20° C); a weak wind; a stable temperature inversion within the atmospheric boundary layer (at altitudes 880–1200 m above ground level); atmospheric pressure: 970–920 hPa. These stable

http://dx.doi.org/10.5772/65274

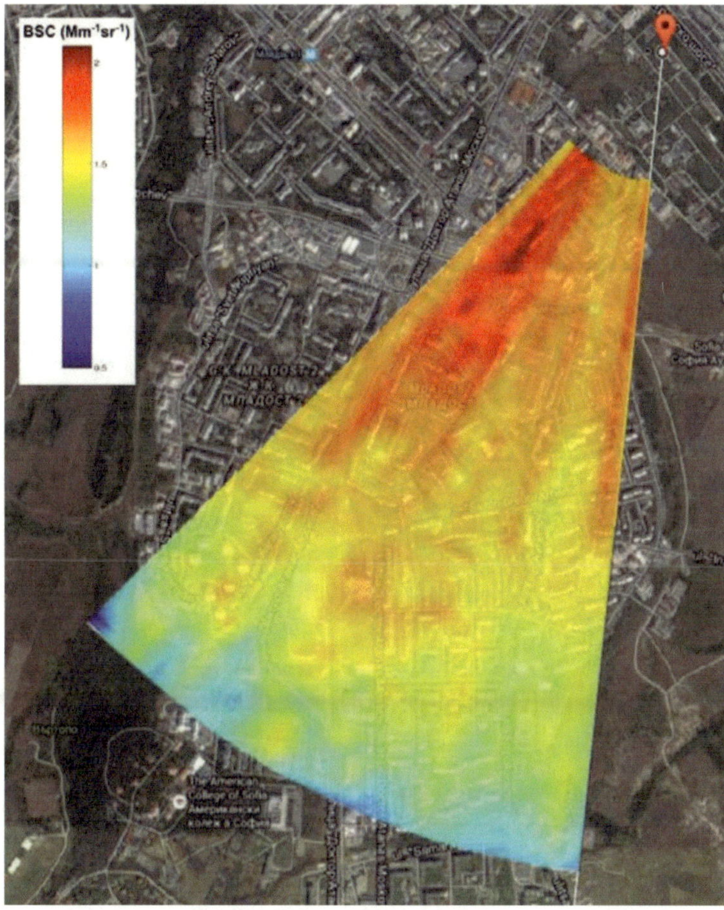

Figure 10. Colormap of the near-surface aerosol density distribution, as measured in the time interval 12:00–14:50 LT on 3 November 2015, in an azimuth range of 40° and distances of up to 3 km.

conditions, in combination with the absence of specific aerosol loadings (e.g. fire smoke, desert dust, etc.) in that period, resulted in aerosol distribution pictures generally similar to those obtained from the individual measurements conducted. Still, the lidar data exhibit particular patterns of the aerosol fields above the city, determined by various local horizontal and vertical air circulations in the close-to-the surface atmospheric layer.

Three separate lidar measurements were carried out on 3 November 2015—one in the first half of the day and two successive ones in the evening. During the daytime measurement, the lidar scanning was performed in south-southwest directions within a horizontal angle range of 40° to distances of up to 3 km. The results are shown in **Figure 10**. The correspondence between the aerosol BCS values and the lidar map colors is given by the color bar in the upper left corner. Inhomogeneous distribution of the aerosol concentration was registered, according to the spotted colormap pattern. In the left-hand upper part of the map, a dark-red colored area can be seen, extending to about 1 km and corresponding to the highest aerosol loading. This

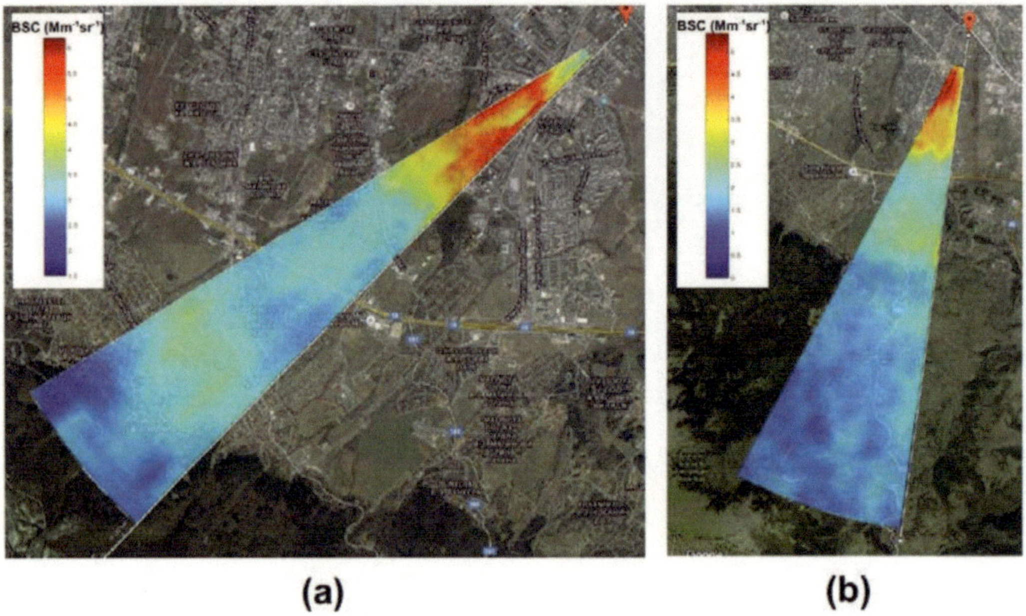

Figure 11. Colormap of the aerosol density distribution as measured on 3 November 2015 in the time interval 19:05–19:41 LT at distances of up to 8 km (a) and in the interval 19:56–20:40 LT at distances up to 12 km (b).

observation is reasonable, taking into account the fact that in this part of the city densely populated residential districts are located, with intense daytime street traffic.

In the evening of 3 November 2015, two successive lidar soundings were performed, the results of which are presented in **Figure 11**. Juxtaposing data of such successive measurements conducted in the same angular sector allows one to follow temporal variations of the aerosol fields over the areas investigated.

The first measurement was carried out by horizontal lidar scanning in an angular sector of 14°, reaching distances of up to 8 km (**Figure 11(a)**), whereas the second one, in a sector of 20° to a distance of 12 km (**Figure 11(b)**). The larger distances reached during the evening measurements are due to the much lower optical background than the daytime one. The comparison of the daytime (**Figure 10**) and nighttime (**Figure 11(a)**) lidar soundings showed that the relatively high concentration of aerosols measured at midday over the zone to 3 km near the lidar was preserved until the evening. At longer distances (beyond the ring road), approaching the Vitosha Mountain, the aerosol concentration decreased and remained relatively homogeneous, as indicated by the low-contrast light-bluish coloring of the corresponding map parts.

As an exception, an increased aerosol density could be observed over some remote parts of the scanned region in the distance range 4–6 km (colored in light green-yellow), where residential districts are located in the mountain skirts. This can be ascribed either to the presence of a light fog or to smoke emissions taking into account the started heating season.

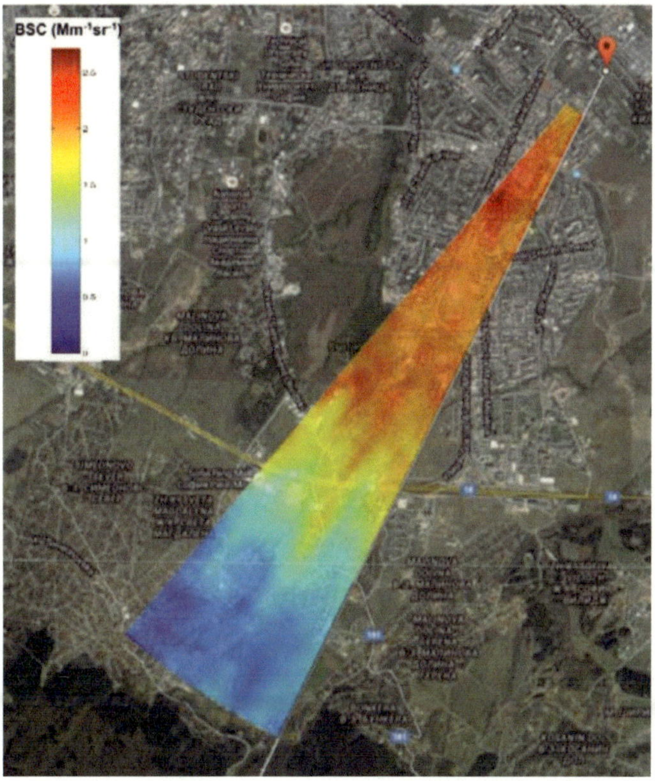

Figure 12. Colormap of the aerosol density distribution as measured on 5 November 2015 in the time interval 12:26–13:25 LT at distances of up to 7 km.

Four measurements were conducted on 5 November 2015 – one daytime over distances of up to 4 km and three successive nighttime ones over distances of up to 11 km. The daytime measurement was carried out in an angular sector of 14° and distances of up to 7 km. The highest aerosol loading was observed above the city zone about 4 km away from the lidar station, reaching the ring road, with a relatively homogeneous aerosol density distribution (**Figure 12**). These results are comparable to the ones presented above obtained during the daytime lidar measurements performed on 3 and 4 November 2015. This is reasonable because of the similar meteorological conditions and the absence of unusual aerosol pollutions.

All three nighttime lidar measurements took place in successive 1 h time intervals, over the same area in an angular sector of 30° and distances up to 12 km. The results are presented as colormaps in **Figure 13**. The comparison of the three panels shows the disappearance of the dynamic atmospheric processes, resulting in a considerable redistribution of the near-surface aerosol density. This can be clearly seen in the figure panels as expressed by the variable color pattern of the maps, most evident in the zones near the lidar extending to 4–5 km (colored mainly in red and yellow). In addition, the extension observed of the blue-colored remote part of the colormap sectors to the city zone in the course of the measurements could be ascribed to

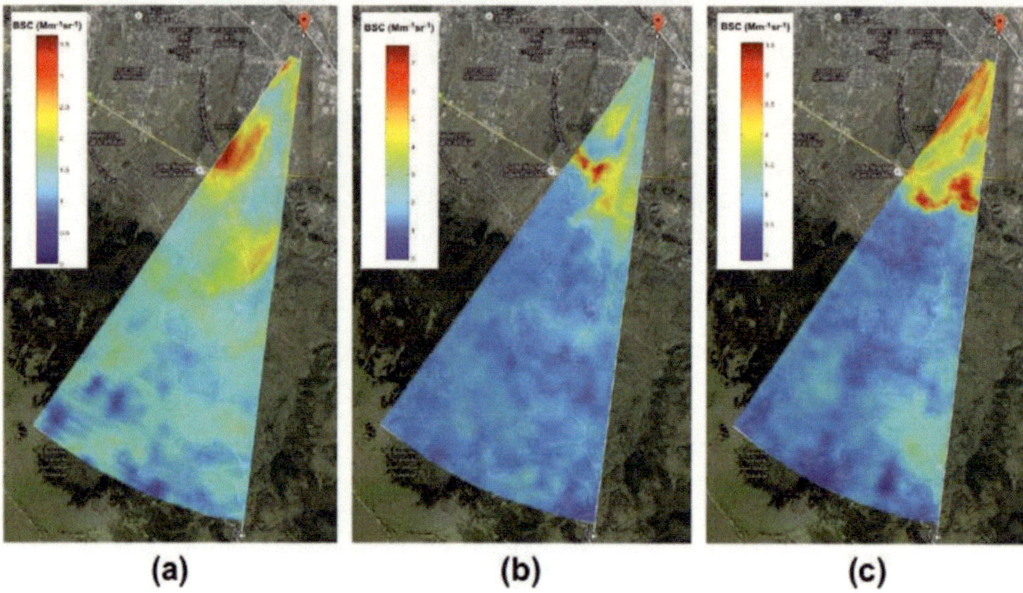

Figure 13. Colormap of the aerosol density distribution as measured on 5 November 2015 in the time intervals 18:37–19:33 LT (a), 19:36–20:29 LT (b) and 20:35–21:28 LT (c), at distances of up to 12 km.

movements of deficient in aerosols air masses from the mountain areas to the city, driven by the evening mountain breeze—characteristic of the Sofia region.

On 6 November 2015, three (two daytime and one nighttime) mapping lidar measurements were carried out within an angular sector of 26° over the same area. The two daytime measurements

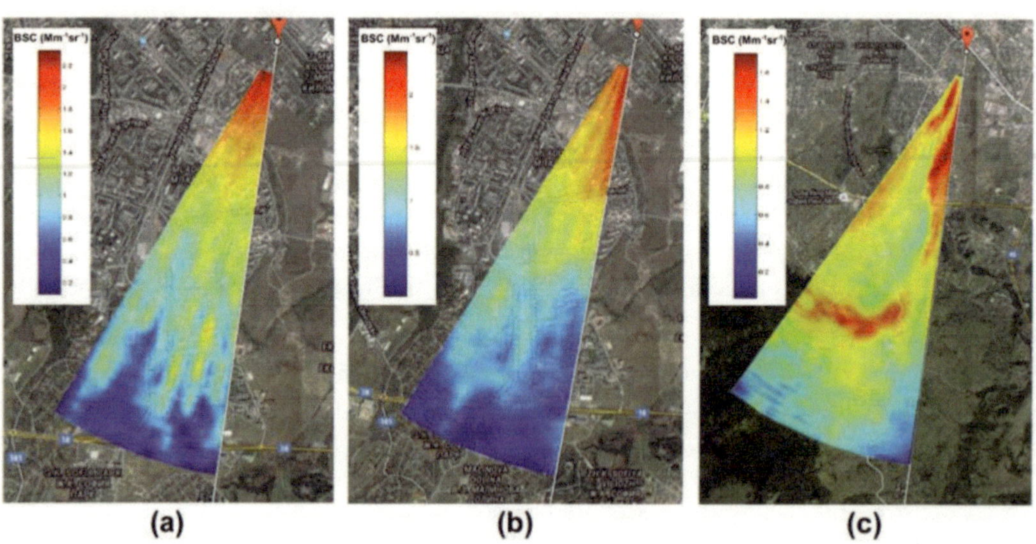

Figure 14. Colormap of the aerosol density distribution as measured on 6 November 2015 in the time intervals 10:24–11:39 LT (a), 11:48–11:53 LT (b) and 18:23–19:19 LT (c), at distances of up to 4 km (a, b) and 10 km (c).

http://dx.doi.org/10.5772/65274

covered distances of up to about 4 km, whereas the evening one, to nearly 10 km. The corresponding results are displayed as colormaps on the three panels of **Figure 14**. The analysis of the two daytime lidar maps (**Figure 14(a)** and **(b)**) revealed the highest aerosol densities in the 1 km zone near the lidar station (colored in red-yellow in the map). Examining the map color pattern, one can perceive gradations of coloring from dominating red in the near zone, through yellow-green in the middle part (forming stripe-like structures), to mainly blue approaching the mountain zone. This grading pattern might be explained as resulting from the action of air currents moving from the city to the mountain, capturing and transporting urban/anthropogenic aerosols to the mountain areas. The lidar data obtained during the nighttime mapping scans also support such an explanation. The colormap in this last case (**Figure 14(c)**) shows a particular aerosol distribution structure dominated by a folded aerosol plume (colored mainly in red and red-yellow) extending from the close-to-the lidar city zone, through the suburbs, up to the mountain skirts, and consisting of two differentiated but connected parts—a dense aerosol field over the urban zones and a similar one at the plane-mountain interface zone. This picture illustrates the complex nature and variability of the near-surface aerosol distribution and spreading, originating from different natural and anthropogenic sources and driven in a complex manner by the local air circulation system.

4. Synergy-based approach for incorporation of the lidar near-surface maps into modern air-quality monitoring systems

As shown in the previous sections, the aerosol lidar mapping technique is capable of providing a fast, accurate, and reliable range-time-resolved determination of optical parameters of the near-surface aerosols, such as the extinction and backscattering coefficients (directly proportional to the aerosol mass concentration), covering broad observation areas. In order to achieve complete quantitative aerosol characterization, determination of the aerosol mass concentration itself is also required. On the other hand, the existing set of in-situ air-pollution detectors present at some sites in the city is able to determine the aerosol mass concentration. However, this is possible to be done just for a limited number of detector location points. We consider that, by combining the two mentioned approaches, particularly by using in-situ obtained data to calibrate the aerosol lidar measurements, a synergistic effect could be achieved, allowing direct mapping of the aerosol mass concentration over the whole urban area. Below, we analyze and discuss the possibilities of achieving such a synergy in the characterization of near-surface aerosol pollutions.

As is well known, the typical existing air monitoring city systems contain the following basic structural components: (1) a network of a limited number of in-situ aerosol, gas, and biological sensors; (2) a network of meteorological sensors; (3) a modeling and data-processing system. The use of a low number of sensors by two networks over large urban areas imposes serious limitations on the information capabilities of the air-quality systems. The lidar maps can be considered as being a (virtual) aerosol sensor network of closely distributed very large number of single aerosol sensor cells of dimensions determined by the lidar maps' spatial resolution. Therefore, the lidar mapping of near-surface aerosol fields appears to be a promising technology for improving the information quality of air-monitoring systems. The combination of the three sensor networks mentioned above incorporated in a joint air-quality monitoring system would provide a synergistic aerosol characterization.

Finally, we have to note that aerosol lidar maps could provide large amounts of additional information about the near-surface atmosphere. They contain data on the near-surface dynamics of air masses, driven by the surface winds but affected by the city structures. This is an important view of a better evaluation of the pollution transport over an urban area. Also, applying multiwavelength lidar mapping, one could contribute to the characterization of the aerosols' size parameters, as well as to identifying their types and origin.

5. Conclusions

Summarizing the results of lidar measurements presented here, one can draw the following conclusions:

1. Lidar mapping of near-surface aerosols, based on (quasi)horizontal and vertical lidar scanning using the two lidar systems of the LRL of IE-BAS, appears to be an effective approach to the accurate and reliable determination of the density, spatial distribution, and temporal dynamics of close-to-ground aerosols, covering broad urban areas in Sofia region.
2. The analysis of the two-dimensional aerosol lidar maps obtained, as superposed on the topological map of Sofia region, shows a good correlation between the aerosol density distribution and the locations of important sources of aerosol pollutions in the zones of observation, such as city streets with intense traffic, densely populated areas, etc.
3. Combining the near-surface aerosol lidar mapping technology with the existing set of in-situ air-pollution detectors and related numerical models and computing facilities, is shown to be a promising synergistic approach to the development of more efficient modern city air-quality monitoring systems.

Acknowledgments

The financial support of the Municipality of Sofia is gratefully acknowledged. The support for EARLINET in the ACTRIS Research Infrastructure Project by the European Union's Horizon 2020 research and innovation program under grant agreement no. 654169 and previously under grant agreement no. 262254 in the 7th Framework Programme (FP7/2007–2013) is also acknowledged.

Author details

Tanja Dreischuh*, Ivan Grigorov, Zahary Peshev, Atanaska Deleva, Georgi Kolarov and Dimitar Stoyanov

*Address all correspondence to: tanjad@ie.bas.bg

Institute of Electronics, Bulgarian Academy of Sciences, Sofia, Bulgaria

References

[1] European Environment Agency (EEA). The European Environment. State and Outlook 2010. Air Pollution. Luxembourg: Publication Office of the European Union; 2010. 46 p. DOI: 10.2800/57792

[2] Fuzzi S, et al. Particulate matter, air quality and climate: lessons learned and future needs. Atmos. Chem. Phys. 2015;**15**:8217–8299. DOI: 10.5194/acp-15-8217-2015

[3] World Health Organisation (WHO). Review of evidence on health aspects of air pollution —REVIHAAP Project. Technical Report. Copenhagen, Denmark: World Health Organisation, WHO Regional Office for Europe; 2013.

[4] Cesaroni G, et al. Long term exposure to ambient air pollution and incidence of acute coronary events: prospective cohort study and meta-analysis in 11 European cohorts from the ESCAPE Project. BMJ (Clinical research ed.). 2014;**348**:f7412. DOI: 10.1136/bmj.f7412

[5] Straif K, Cohen A, Samet J, editors. Air Pollution and Cancer. IARC Scientific Publication. Geneva, Switzerland: World Health Organisation; 2013. 161 p.

[6] Birmili W, et al. Atmospheric aerosol measurements in the German Ultrafine Aerosol Network (GUAN)—Part 3: Black Carbon mass and particle number concentrations 2009 to 2014. Gefahrstoffe Reinhaltung der Luft (Air Pollution Control). 2015;**75**:479–488.

[7] Sicard M, et al. EARLINET: potential operationality of a research network. Atmos. Meas. Tech. 2015;**8**:4587–4613. DOI: 10.5194/amt-8-4587-2015

[8] Kovalev V, Eichinger W. Elastic Lidar: Theory, Practice, and Analysis Methods. Hoboken, NJ, USA: Wiley-Interscience; 2004. 615 p. DOI: 10.1002/0471643173

[9] Pappalardo G, et al. EARLINET: towards an advanced sustainable European aerosol lidar network. Atmos. Meas. Tech. 2014;**7**:2389–2409. DOI: 10.5194/amt-7-2389-2014

[10] Shan J, Toth C, editors. Topographic Laser Ranging and Scanning: Principles and Processing. Boca Raton, FL, USA: Taylor and Francis; 2009. 590 p. DOI: 10.1080/0143116 0903112612

[11] Werner Ch. Doppler wind lidar. In: Weitkamp C, editor. Lidar. Range-Resolved Optical Remote Sensing of the Atmosphere. New York: Springer; 2005. pp. 325–354. DOI: 10.1007/b106786

[12] Pichugina Y, Banta R, Brewer W, Sandberg S, Hardesty R. Doppler Lidar-based wind-profile measurement system for offshore wind-energy and other marine boundary layer applications. J. Appl. Meteor. Climatol. 2012;**51**:327–349. DOI: 10.1175/JAMC-D-11-040.1

[13] Radlach M, Behrendt A, Wulfmeyer V. Scanning rotational Raman lidar at 355 nm for the measurement of tropospheric temperature fields. Atmos. Chem. Phys. 2008;**8**:159–169. DOI: 10.5194/acp-8-159-2008

[14] Behrendt A, Pal S, Wulfmeyer V, Valdebenito B. A-M, Lammel G. A novel approach for the characterization of transport and optical properties of aerosol particles near sources —Part I: measurement of particle backscatter coefficient maps with a scanning UV lidar. Atmos. Environ. 2011;**45**:2795–2802. DOI: 10.1016/j.atmosenv.2011.02.061

[15] Spuler S M, Mayor Sh D. Eye-safe aerosol lidar at 1.5 microns: progress towards a scanning lidar network. Proc. SPIE. 2007;**6681**:668102. DOI: 10.1117/12.739519

[16] Gao F, Bergant K, Filipčič A, Forte B, Hua D-X, Song X-Q, Stanič S, Veberič D, Zavrtanik M. Observations of the atmospheric boundary layer across the land–sea transition zone using a scanning Mie lidar. J. Quant. Spectrosc. Radiat. Trans. 2011;**112**:182–188. DOI: 10.1016/j.jqsrt.2010.04.001

[17] Xie C, Zhao M, Wang B, Zhong Z, Wang L, Liu D, Wang Y. Study of the scanning lidar on the atmospheric detection. J. Quant. Spectrosc. Radiat. Trans. 2015;**150**:114–120. DOI: 10.1016/j.jqsrt.2014.08.023

[18] Stoyanov D, Grigorov I, Kolarov G, Peshev Z, Dreischuh T. LIDAR atmospheric sensing by metal vapor and Nd:YAG lasers. In: Fadhali M, editor. Advanced Photonic Sciences. Rijeka, Croatia: Intech; 2012. pp. 345–374. DOI: 10.5772/29325

[19] Stoyanov D, Dreischuh T, Grigorov I, Kolarov G, Deleva A, Peshev Z, Nedkov I. Aerosol lidar mapping of large urban areas over Sofia Municipality. On the synergy with in-situ atmospheric sensors. Abstracts of the papers presented in the Fifth Scientific Meeting of the COST Action TD1105—New Sensing Technologies for Air-Pollution Control and Environmental Sustainability; 16–18 December 2015; Sofia, Bulgaria.

[20] He T-Y, Stanič S, Gao F, Bergant K, Veberič D, Song X-Q, Dolžan A. Tracking of urban aerosols using combined LIDAR-based remote sensing and ground-based measurements. Atmos. Meas. Tech. 2012;**5**:891–900. DOI: 10.5194/amt-5-891-2012

[21] Basart S, Pay M, Jorba O, Perez C, Jimenez-Guerrero P, Schulz M, Baldasano J. Aerosols in the CALIOPE air quality modelling system: evaluation and analysis of PM levels, optical depths and chemical composition over Europe. Atmos. Chem. Phys. 2012;**12**:3363–3392. DOI: 10.5194/acp-12-3363-2012

[22] Peshev Z, Dreischuh T, Toncheva E, Stoyanov D. Two-wavelength lidar characterization of atmospheric aerosol fields at low altitudes over heterogeneous terrain. J. Appl. Remote Sens. 2012;**6**:063581. DOI: 10.1117/1.JRS.6.063581

[23] Deleva A, Grigorov I. Lower troposphere observation over urban area with lidar at 1064 nm. IJNO. 2011;**2011**:1–8. DOI: 10.1155/2011/769264

[24] Peshev Z, Evgenieva Ts, Dreischuh T, Stoyanov D. Two-wavelength lidar characterization of optical, dynamical, and microphysical properties of Saharan dust layers over Sofia, Bulgaria. Proc. SPIE. 2015;**9447**:94470R. DOI: 10.1117/12.2175654

[25] Grigorov I, Stoyanov D, Kolarov G. Lidar observation of volcanic dust layers over Sofia. Proc. SPIE. 2011;**7747**:77470R. DOI: 10.1117/12.882795

[26] Grigorov I, Deleva A, Stoyanov D, Kolev N, Kolarov G. LIDAR detection of forest fire smoke above Sofia. Proc. SPIE. **7747**:77470U. DOI: 10.1117/12.2178791

[27] Measures R. Laser Remote Sensing: Fundamentals and Applications. New York: Wiley-Interscience; 1984; p. 510.

[28] Klett, J. Stable analytical inversion solution for processing lidar returns. Appl. Opt. 1981;**20**:211–220. DOI: 10.1364/AO.20.000211

[29] Fernald G. Analysis of atmospheric LiDAR observations: some comments. Appl. Opt. 1984;**23**:652–653. DOI: 10.1364/AO.23.000652

[30] United States Committee on Extension to the Standard Atmosphere (COESA). U.S. Standard Atmosphere 1976. U.S. Government Printing Office, Washington, D.C. 1976. 241 p.

[31] Ansmann A, Seifert P, Tesche M, Wandinger U. Profiling of fine and coarse particle mass: case studies of Saharan dust and Eyjafjallajökull/Grimsvötn volcanic plumes. Atmos. Chem. Phys. 2012;**12**:9399–9415. DOI: 10.5194/acp-12-9399-2012

Chapter 5

Aerosols Monitored by Satellite Remote Sensing

Zhihua Mao, Xueliang Deng, Peng Chen,
Bangyi Tao, Guanying Yang, Yanfeng Huo and
Qiankun Zhu

Additional information is available at the end of the chapter

http://dx.doi.org/10.5772/65284

Abstract

Aerosols, small particles suspended in the atmosphere, affect the air quality and climate change. Their distributions can be monitored by satellite remote sensing. Many images of aerosol properties are available from websites as the by-products of the atmospheric correction of the satellite data. Their qualities depend on the accuracy of the atmospheric correction algorithms. The approaches of the atmospheric correction for land and ocean are different due to the large difference of the ground reflectance between land and ocean. A unified atmospheric correction (UAC) approach is developed to improve the accuracy of aerosol products over land, similar to those over ocean. This approach is developed to estimate the aerosol scattering reflectance from satellite data based on a lookup table (LUT) of in situ measured ground reflectance. The results show that the aerosol scattering reflectance can be completely separated from the satellite measured radiance over turbid waters and lands. The accuracy is validated with the mean relative errors of 22.1%. The vertical structures of the aerosols provide a new insight into the role of aerosols in regulating Earth's weather, climate, and air quality.

Keywords: aerosol, epsilon, aerosol optical thickness, Angstrom exponent, atmospheric correction, land remote sensing, ocean colour remote sensing, the UAC model

1. Introduction

Aerosols, suspended particulate matter in air, act as a crucial factor in global climatic fluctuations [1]. Aerosols can affect the climate through absorption and scattering of solar radiation [2] and therefore perturb the radiation budget and contribute to radiative forcing [3]. Aerosols may change the size and density of cloud droplets, thus modify the cloud albedo, the cloud lifetime and the precipitation [4]. Aerosols also influence air quality and therefore affect human health [5]. Current uncertainties of aerosols in the Earth radiation budget limit our understanding of the climate system and the potential for global climate change [1]. Satellite observations are needed to understand the distribution and impact of aerosols on regional and global scales [6]. Satellites can monitor some aerosol optical properties, e.g. aerosol optical thickness (AOT) and Angstrom exponent, the key factors for climate change research [7]. In fact, these properties can be retrieved during the atmospheric correction of satellite images.

Ever since Gordon [8] designed an atmospheric correction approach based on the black ocean assumption (BOA) at two near-infrared (NIR) bands, this approach has been widely applied to process satellite ocean colour remote sensing data. The performance of the approach was improved significantly [9, 10]. However, this approach still faces problems in Case 2 waters [11]. Some other algorithms of atmospheric corrections have been developed especially for the coastal waters. These include the use of the assumption of spatial homogeneity of the NIR band ratio [12], the spectral shape matching methods [13], an iterative fitting algorithm with the bio-optical models [14], the BOA method using the short wave infrared (SWIR) bands over turbid waters [15] or the algorithm using the ultraviolet bands [16].

The atmospheric correction over land meets more complicated situations. Similar to the BOA approach for the ocean colour remote sensing, the dark target (DT) approach has been widely used to estimate the optical properties of aerosols over land [17]. Other approaches have been developed using different methods, for example, the invariant object approach by Hall et al. [18], the histogram matching by Richter [19] and the radiative transfer model by Gao et al. [20]. Traditionally, different approaches of the atmospheric correction are necessary for land and ocean to optimize each case. Recently, Mao et al. [21, 22] developed an approach to estimate the aerosol scattering reflectance over turbid waters based on a look-up table (LUT) of in situ measurements. Following this approach, a unified atmospheric correction (UAC) approach is developed for both land and ocean [23].

Over the last several decades, satellite remote sensing has provided an increasingly detailed view of aerosols and clouds [24] but limited with column-averaged aerosol properties. Aerosols in the lowest part of the atmosphere are likely to be removed quickly by the rain, those in higher altitudes are much more likely to travel long distances and affect air quality in distant regions. The Cloud-Aerosol Lidar and infrared pathfinder satellite observation (CALIPSO) satellite provides new capabilities to distinguish aerosol optically thin boundary layer from cloud by considering the vertical thickness and location of the layers as well as from the spectral behaviour of the lidar backscatter [25], useful in studying the interactions between aerosols and clouds with their roles in the climate system.

2. Methods

2.1. Retrieving the aerosols based on the BOA method

In the atmospheric correction procedure, the aerosol scattering reflectance needs to be estimated from remote sensing data, relying on the condition that it can be clearly separated from the total satellite-measured reflectance. This condition can be met using the BOA method over clear oceanic waters when the water-leaving reflectance in the NIR bands is negligible. The satellite-received reflectance at the TOA was defined by:

$$\rho_t(\lambda) = \pi L_t(\lambda) / (F_0(\lambda)\cos\theta_0) \tag{1}$$

where $L_t(\lambda)$ is the satellite-measured radiance, $F_0(\lambda)$ is the extra-terrestrial solar irradiance, and θ_0 is the solar-zenith angle. Wang [26] partitioned the term $\rho_t(\lambda)$ into components corresponding to distinct physical processes by:

$$\rho_t(\lambda) = \rho_r(\lambda) + \rho_A(\lambda) + t(\lambda)\rho_{wc}(\lambda) + T(\lambda)\rho_g(\lambda) + t(\lambda)\rho_w(\lambda) \tag{2}$$

where $\rho_r(\lambda)$ is the Rayleigh scattering reflectance due to the air molecules, $\rho_A(\lambda)$ is the aerosol scattering reflectance including the Rayleigh-aerosol interactions, $\rho_{wc}(\lambda)$ is the reflectance of the ocean whitecaps, $\rho_g(\lambda)$ is the reflectance of Sun glitter off the sea surface and $\rho_w(\lambda)$ is the water-leaving reflectance. The terms $t(\lambda)$ and $T(\lambda)$ are the diffuse and direct transmittances of the atmosphere, respectively.

$$\rho_A(\lambda) = \rho_t(\lambda)\text{-}\rho_r(\lambda) - T(\lambda)\rho_g(\lambda) - t(\lambda)\rho_{wc}(\lambda) - t(\lambda)\rho_w(\lambda) \tag{3}$$

When the water-leaving reflectance in the two NIR bands is assumed to be zero, the aerosol scattering reflectance can be obtained and used to compute the aerosol single scattering reflectance from:

$$\rho_{AS}(\lambda) = \frac{-b + \sqrt{b^2 - 4c(a - \rho_A(\lambda))}}{2c} \tag{4}$$

The epsilon spectrum of the aerosol single scattering reflectance is then defined as follows:

$$\varepsilon(\lambda_i, \lambda_0) = \frac{\rho_{AS}(\lambda_i)}{\rho_{AS}(\lambda_0)} \tag{5}$$

Angstrom exponent n is computed from:

$$n = \frac{\lambda_i}{\lambda_0} \cdot \varepsilon(\lambda_i, \lambda_0) \tag{6}$$

According to Ref. [27], the aerosol optical thickness (AOT) can be calculated from the aerosol single scattering reflectance as follows:

$$\tau_a(\lambda) = \rho_{AS}(\lambda) \Big/ \omega_a(\lambda) P_a(\lambda) \tag{7}$$

where $\omega_a(\lambda)$ is the aerosol single scattering albedo and $P_a(\lambda)$ is the aerosol scattering phase function.

2.2. Retrieving the aerosols based on the UAC model

The assumption of the BOA method becomes invalid over turbid waters and lands, leading to the failure of the standard atmospheric correction. A new approach needs to be developed to retrieve the aerosols from the satellite data. We define $\rho_{AW}(\lambda)$ as a term of the aerosol-water reflectance, which includes aerosol scattering reflectance and the ground reflectance at the top of the atmosphere (TOA), derived as follows.

$$\rho_{AW}(\lambda) = \rho_t(\lambda) \text{-} \rho_r(\lambda) - T(\lambda)\rho_g(\lambda) - t(\lambda)\rho_{wc}(\lambda) \tag{8}$$

Then, aerosol scattering reflectance $\rho_A(\lambda)$ can be obtained using:

$$\rho_A(\lambda) = \rho_{AW}(\lambda) - t(\lambda) \cdot \rho_w(\lambda) \tag{9}$$

It is derived from the normalized water-leaving reflectance, which is selected from a lookup table of the in situ measurements using the UAC method.

The epsilon spectrum is used to match the two closest aerosol models and obtain the corrected epsilon values $\varepsilon^c(\lambda_i, \lambda_0)$ following the approach of Mao et al. (2013). The Angstrom exponent $\eta(\lambda_i)$ is obtained from:

$$\eta(\lambda_i) = \ln(\varepsilon^c(\lambda_i, \lambda_0)) \cdot \lambda_0 / \lambda_i \tag{10}$$

We defined λ_M as a new reference wavelength obtained from the mean value of the band wavelengths.

$$\lambda_M = \sum \lambda_i / n \tag{11}$$

http://dx.doi.org/10.5772/65284

The corrected epsilon spectrum is adjusted by a new reference wavelength, defined as follows:

$$\varepsilon_M(\lambda_i, \lambda_M) = \left(\frac{\lambda_M}{\lambda_i}\right)^{\eta(\lambda_i)} \tag{12}$$

The aerosol single scattering reflectance is used to obtain the mean value, defined as follows:

$$\rho_{AS}(\lambda_M) = \sum \rho_{AS}(\lambda_i) / n \tag{13}$$

Then, a new aerosol single scattering reflectance is obtained from:

$$\rho^n{}_{AS}(\lambda_i) = \varepsilon_M(\lambda_i, \lambda_M) \cdot \rho_{AS}(\lambda_M) \tag{14}$$

This approach relies on the assumption that the aerosol scattering reflectance follows the Angstrom law instead of the BOA. This new assumption eliminates the effects of non-zero water-leaving reflectance in the atmospheric correction procedure. Due to the amplifying effects of the epsilon spectrum in estimating the aerosol scattering reflectance, the water-leaving reflectance may be estimated with large errors using the BOA method. One advantage of the UAC model is that the epsilon spectrum can be obtained to find the aerosol model, instead one value in the NIR band. The other advantage is that the difference of the ground reflectance among different objects is represented by different spectra in the LUT. Therefore, the LUT of the ground reflectance helps the UAC model to eliminate the different effects of the atmospheric correction due to the large difference of the reflectance between land and ocean.

2.3. Comparison between the UAC model and the BOA method

It is well known that the processing of satellite data using the standard atmospheric correction usually fails over turbid coastal regions. One of the main reasons is the difficulty in accurately determining the epsilon values. Epsilon is used to determine the magnitude of the aerosol scattering reflectance and deficiencies in its determination degrades the accuracy of the atmospheric correction of satellite remote sensing data. Normally, epsilon is estimated from aerosol scattering reflectance under the assumption of zero water-leaving radiance in the two NIR bands. However, the water-leaving radiance values in the two NIR bands are usually much higher than zero in the coastal waters [28], causing errors in the epsilon estimation.

This overestimation is caused by that the water-leaving reflectance is wrongly attributed as part of the aerosol scattering reflectance to overestimate the epsilon value. A small bias of the epsilon may easily lead to a relatively large error in the water-leaving reflectance due to the amplifying effects of the extrapolation of epsilon. The aerosol scattering reflectance obtained by the UAC model is different from that by the BOA method, especially in the coast regions.

A simple comparison is made based on three selected typical locations (marked by the green crosses in **Figure 3**), and the results are shown in **Figure 1**.

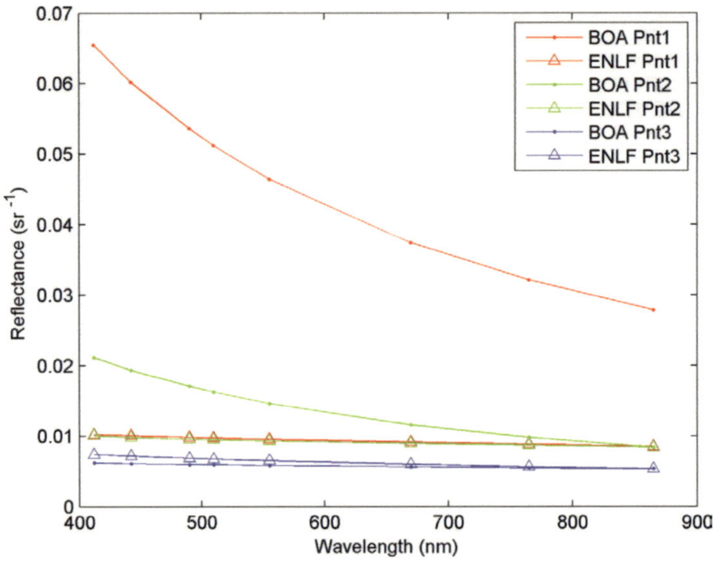

Figure 1. Comparison between the aerosol scattering reflectance using the UAC and the BOA method.

Epsilon	Band 1	Band 2	Band 3	Band 4	Band 5	Band 6	Band 7	Band 8
1–1.04	8.94	8.44	7.72	7.42	6.74	5.06	4.26	4.51
1.04–1.07	11.77	11.21	10.41	10.09	9.37	7.80	7.03	6.86
1.07–1.1	74.38	70.16	64.40	62.17	57.49	47.43	41.04	37.29
1.1–1.2	206.63	186.90	161.39	151.86	132.73	94.93	72.27	55.23
1.2–1.3	453.50	394.95	323.45	297.96	248.86	159.89	111.68	75.14
>1.3	961.04	796.23	612.95	552.07	441.02	259.90	172.08	110.42

Table 1. The difference between the aerosol reflectance obtained from the BOA method and that from the UAC model over the different epsilon ranges.

The aerosol scattering reflectance calculated from the two methods are similar to each other in oceanic waters (e.g. Location 3). The differences of the two reflectance spectra become larger near the coastal regions, with a mean relative error over 50% (e.g. Location 2). The difference of the two aerosol scattering reflectance values is very small in Band 8 and large in Band 1. This difference is attributed to the different epsilon values of the two methods, with 1.17 using the BOA method and 1.03 using the UAC model, respectively. A larger epsilon value will obviously overestimate the aerosol scattering reflectance, but this error is difficult to detect until a negative value of the water-leaving reflectance is produced. The BOA reflectance in Band 1 is

http://dx.doi.org/10.5772/65284

5.4 times larger than the UAC reflectance in highly turbid waters (e.g. Location 1). The value in Band 8 is also overestimated using the BOA method.

To evaluate the difference between the aerosol scattering reflectance using the two methods, the relative errors of the reflectance are computed according to the epsilon ranges (**Table 1**), in which the values using the UAC model are taken as the truth data.

From **Table 1**, the relative differences of the aerosol reflectance from the BOA method and the UAC model vary largely over the different epsilon ranges. The differences in all bands are relatively small for the epsilon value range of 1–1.07. The differences become higher than 50% for the range of 1.07–1.1. The differences become too much large when epsilon values are higher than 1.1, with a mean difference of 488% for the range of >1.3. Therefore, the BOA method is valid only with the epsilon value of less than 1.07.

A small error of the reflectance in NIR bands will cause somehow relatively larger error of the aerosol scattering reflectance using the BOA method. For example, two aerosol-water reflectance spectra are selected from two neighbouring pixels and shown in **Figure 2**, together with two aerosol scattering reflectance spectra obtained by the BOA method.

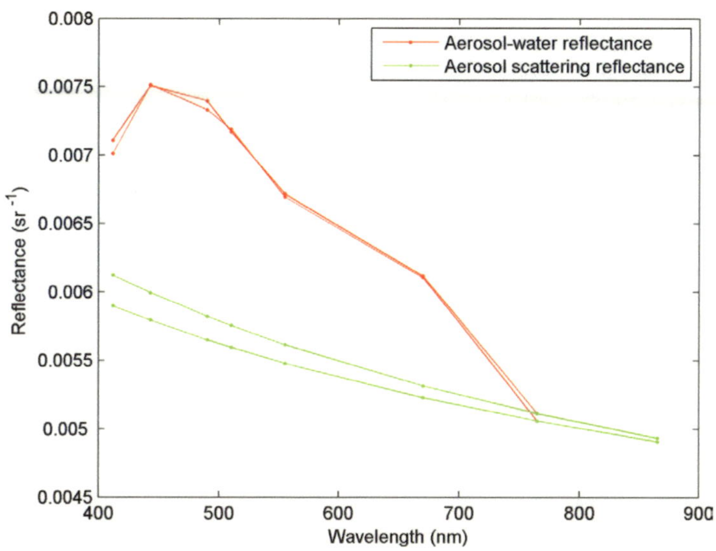

Figure 2. Comparison between the reflectance spectra at two neighbouring pixels. The red lines represent the aerosol-water reflectance from satellite data and the green lines are the aerosol scattering reflectance using the BOA method.

The magnitudes of the two aerosol-water reflectance are close to each other. But the differences of the reflectance in NIR bands lead to obtain two different epsilon values of 1.031 and 1.036 using the BOA method. Due to the amplifying effects, the difference in the two aerosol scattering reflectance is 2.26×10^{-4} sr^{-1} in Band 1, about 10-folds higher than that in Band 8. This difference tends to spread to the water-leaving reflectance in the atmospheric correction procedure with 2.1×10^{-4} sr^{-1} in Band 1, about threefolds of the difference between the satellite reflectance. The difference of the two aerosol scattering reflectance in Band 2 is up to 50-folds higher than the satellite TOA values. Therefore, magnitudes of the aerosol scattering reflectance in the visible bands are very sensitive to the bias of the reflectance in the NIR bands using the BOA method. This problem can be significantly reduced by the UAC model.

3. Aerosols monitored by satellites over the ocean

3.1. Data

Aerosol robotic network (AERONET) is a ground-based optical aerosol monitoring network used to ensure the quality of aerosol products [29]. We obtained the AOT data from the AERONET project web page (http://aeronet.gsfc.nasa.gov). Data quality control is employed to eliminate errors caused by measurement conditions, such as clouds and biomass burning events [30]. There are a total of 554 measurement sites in the world. The site locations are distributed over islands, coastal areas, inland sites and mountains. The AOT was measured during the 2006 winter and 2007 autumn cruises over the ECS using a handheld multi-band sun photometer (MICROTOPS manufactured by the Solar Light Company). A total of 17 stations during the winter cruise, and 18 stations during the autumn cruise were measured with much more measurements during the transit between stations. To compare values between SeaWiFS and AOT, a matched data set was assembled using a time window for the satellite overpass of ±2 h from the in situ measurement. The locations of the matched dataset are shown in **Figure 3**.

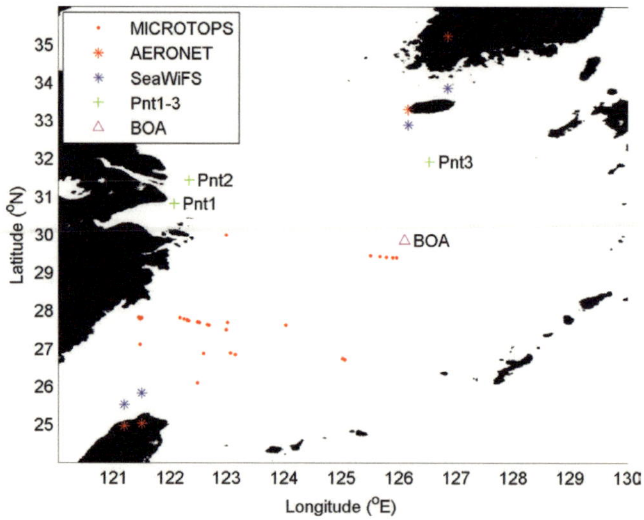

Figure 3. Some locations used in the paper. (1) Red dots represent the locations of matched MICROTOPS measured AOT and SeaWiFS data. (2) The three green crosses are the locations used for the comparison between the UAC model and BOA method. (3) The four blue stars are the locations of the AERONET sites and the nearby blue crosses are the locations for the matched SeaWiFS data. (4) The pink triangle is the location of the SeaWiFS measurements used for the analysis of the BOA method.

3.2. Some results

Normally, the accuracy of the aerosol scattering reflectance depends directly on the value of epsilon. A small error of the epsilon will easily lead to an amplified error in the aerosol scattering reflectance. If the entire epsilon spectrum instead of one value at the NIR band is used to determine the two closest aerosol models, the results are more robust. The key advantage of the UAC model is that it finds suitable aerosol models by matching the epsilon spectrum and thus improves the accuracy of the aerosol reflectance. The UAC model was used to process the SeaWiFS image on March 27, 2007 and the epsilon image is shown in **Figure 4**. To produce this image, the Rayleigh scattering reflectance was computed using a multiple scattering approach to obtain the aerosol-water reflectance. The aerosol scattering reflectance and the epsilon spectra were obtained by the UAC model.

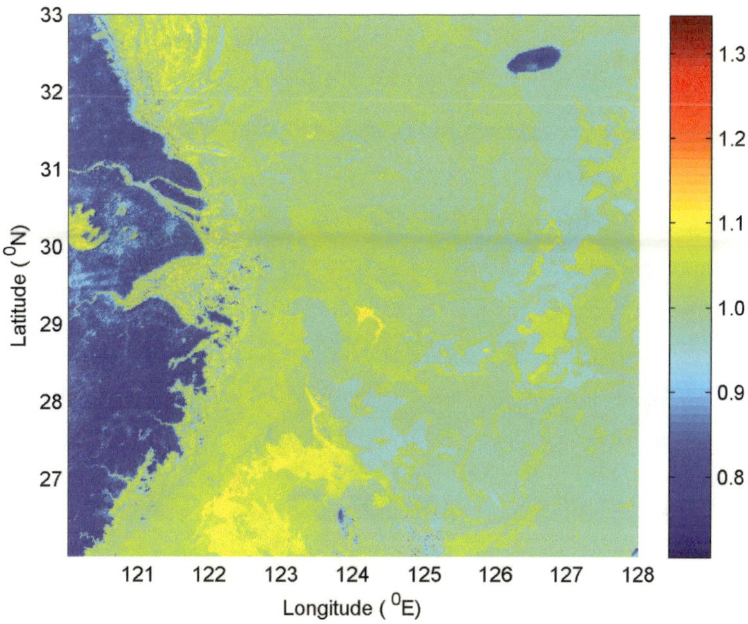

Figure 4. An epsilon image from SeaWiFS data on March 27, 2007 using the UAC model.

From the refined epsilon distribution in **Figure 4**, we can see that the high values of epsilon in the coastal regions are significantly reduced from around 1.4 to 1.05, with values closely in oceanic waters. As the spectra of the reflectance follow the Angstrom law, the structures of images in one band are similar to those in the other bands. The image of the aerosol scattering reflectance in Band 2 is selected and shown in **Figure 5**.

The $\rho_A(\lambda)$ values in Band 2 vary from 0.0042 to 0.015 sr^{-1}, with the mean value of 0.0092 sr^{-1} and the standard deviation of 0.0029 sr^{-1}. The distributions of the image indicate that $\rho_A(\lambda)$ varies significantly from one location to another, with the maxima in the northern part and the minima in the south. From the structures of the image in **Figure 5**, it is clear that the UAC model has almost removed the influence effects of the non-zero water-leaving reflectance in the NIR bands in the coastal regions, even for highly turbid waters.

The water-leaving reflectance in the NIR bands varies from 0 to 0.06 sr^{-1}, almost six times higher than the aerosol reflectance. Under the BOA, up to six times extra values will be falsely added to the actual aerosol reflectance in the NIR bands, which obviously leads to the failure of the atmospheric correction procedure. In comparison, the UAC method makes use of the LUT, the magnitude variation in the ground reflectance has no effects in determining aerosol reflectance. The results show that the UAC model has clearly separated the aerosol scattering reflectance, working well from Case 1 to Case 2 waters.

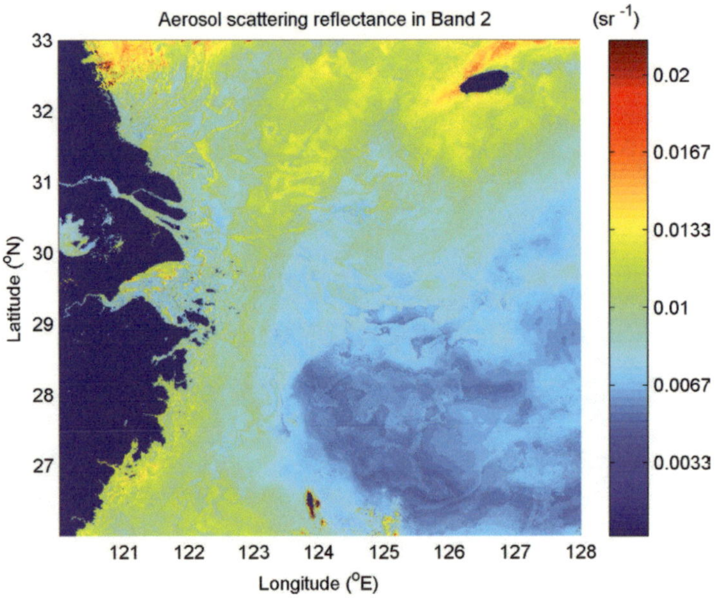

Figure 5. An image of aerosol reflectance retrieved from SeaWiFS Band 2 on March 27, 2007 using the UAC model.

3.3. Validation of the retrieval aerosols over the sea

The globally distributed AERONET data can be used to validate the aerosol products derived from satellite ocean colour remote sensing data such as SeaWiFS [31]. The SeaWiFS reflectance was processed to obtain epsilon using the UAC model, and the results are shown in **Figure 6**. The AOT spectra were used to calculate the aerosol reflectance using the geometric angles of SeaWiFS, to obtain the epsilon values.

Figure 6 shows that the two types of epsilon are relatively comparable. Both the magnitudes and the spectra shapes are similar to each other. AOT derived from MODIS and AERONET must be matched on space and time. Window size on space and time are set as 50 km × 50 km and 1 h in the anterior "http://www.iciba.com/" research. The correlation and regression coefficients show excellent agreement with AERONET measurements over the China Sea.

http://dx.doi.org/10.5772/65284

Meanwhile, the relationship is all good and their correlation coefficients are greater than 0.9. Errors of 65% points are under ± 0.05 ± 0.05 τ based on NASA standard. Fine mode fraction (FMF) is validated over the ocean with the error of about 20%.

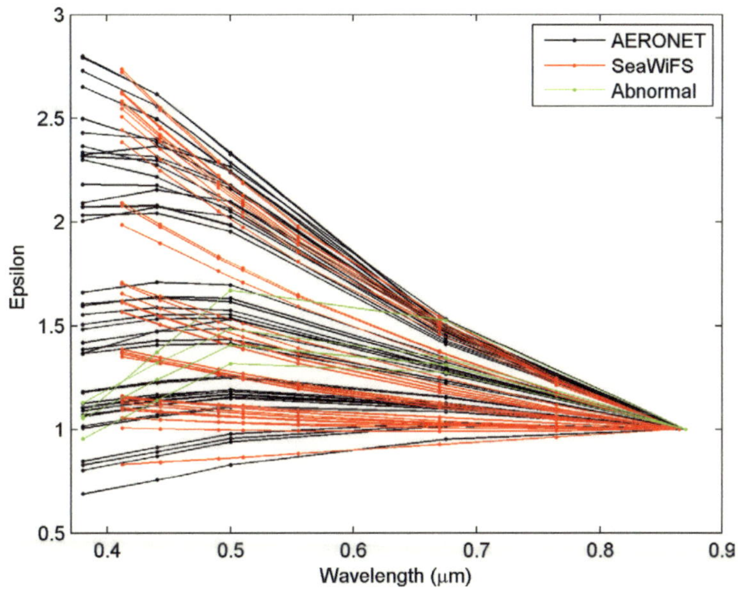

Figure 6. The comparison of the epsilon spectra from SeaWiFS reflectance using the UAC model (Red lines) to those from the AOT measured by the MICROTOPS instrument (Black lines) in the ECS. Green lines represent abnormal epsilon spectra.

Some factors affect the results of the validation. As the AOT data of AERONET are retrieved from the ground based radiance measurements, the data quality of AOT can also be affected by instrument calibration and the measurement conditions [30]. The translation from AOT to the aerosol scattering reflectance is related to the parameters of the aerosol single albedo and its phase function, which are determined by the absorption, and the type and particle size distribution of aerosols. These parameters were not measured, and typical values were used in the computation of epsilon from the AOT.

3.4. Spatial and temporal variations of aerosols over the ocean

The seasonal mean of MODIS AOT over the China Sea from 2001 to 2006, shown in **Figure 7**, demonstrates a notable seasonal change in spatial distribution. The maximum AOT appears in spring with the minimum in summer. In winter, high AOT is distributed along the coast and decreases obviously with the distance far from continent. Maximum AOT has been observed in coastal region and the value of it is larger than 0.5, meanwhile, the minimum AOT appears in open ocean and value of it is 0.1. The distribution of FMF is similar, suggesting that human activity influences aerosols. AOT over the north of 25°N is all larger than 0.17 and it over the south of 25°N is smaller than 0.17. The distribution of AOT in spring is similar to that in winter. Because of dust prevailing, AOT increases obviously and is larger than 0.17 over the whole China Sea. The distribution in summer changes remarkably, with large values in the north and small in the south. Maxima appear over the Yellow Sea and Bohai Sea. Value of AOT in fall becomes the least in four seasons, and the pattern shows similarly structure of that in winter.

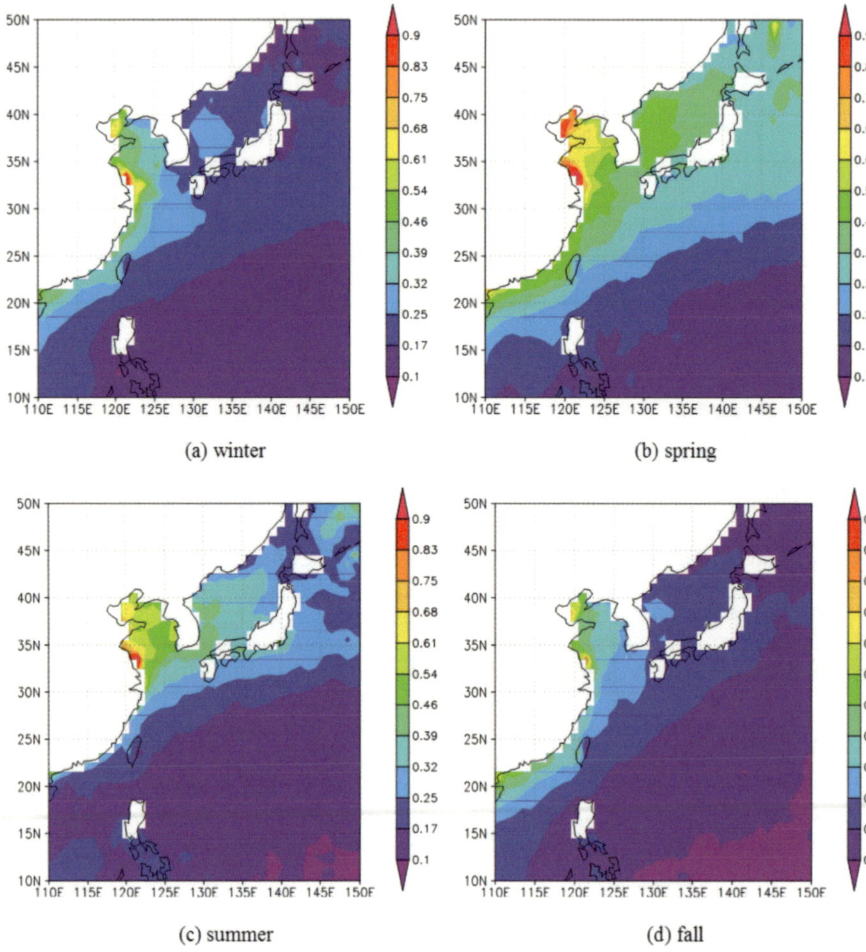

Figure 7. Seasonal distribution of AOT over the China Sea from 2001 to 2006. (a) winters; (b) spring; (c) summer; (d) fall.

Both AOT and FMF exist an annual cycle over the China Sea. The AOT appears maxima in spring with values larger than 0.4 and reaches minimum in summer with values less than 0.25. The FMF also has a periodic oscillation but an opposite tendency. The FMF reaches maxima in summer with values larger than 0.6 and gets minima in spring with values less than 0.5. The mean AOT and FMF during the 6 years (2001–2006) over the China Sea are shown in **Figure 8**. The AOT is the largest in April and the smallest in September, while the

FMF is the largest in September and the smallest in April. Based on the above analysis, we can find that AOT and FMF exist in significant temporal variety with an annual cycle over the China Sea.

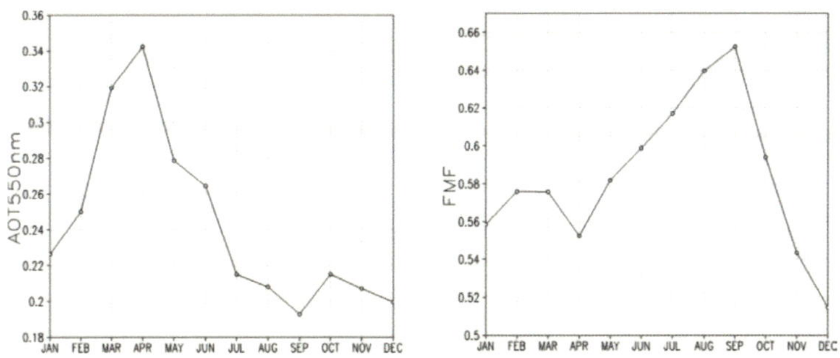

Figure 8. Monthly mean of the AOT (left panel) and the FMF (right panel) over the China Sea.

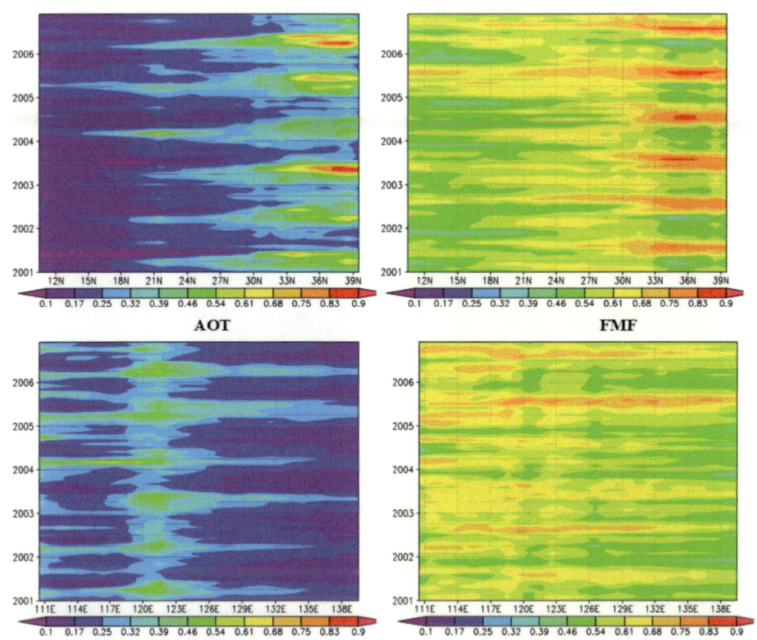

Figure 9. The latitudinal and longitudinal distribution of the AOT and FMF over the China Sea.

How about the spatial distribution of the AOT and FMF over the China Sea? The latitude-time and longitude-time diagrams of the AOT and FMF over the China Sea are shown in **Figure 9**. It represents a spatial characteristic related with the distance far from the coast and important industry regions, due to, which aerosols over the China Sea are mainly influenced by the continent source.

The AOT appears the highest value between 30°N and 40°N in the latitude direction, as there are main industry regions. The FMF also reaches the highest values in the same region. Both AOT and FMF get less towards the southern of the 30°N, due to lack of industry pollution. The human influence is very evident at longitude direction. China coast is around 120°E with the highest of the AOT and the FMF. Both the AOT and FMF gradually decrease away from the coast. The decrease tendency of AOT over the East China Sea has an affinity with the distance far from the continent, suggesting the AOT should be influenced by the continental sources.

3.5. Correlation to meteorological conditions

The aerosol's transport depends on wind, so seasonal change of wind may lead to the spatial and the temporal distribution of aerosols over the China Sea. The mean wind field at 850 hp in 2001–2006 is shown in **Figure 10**. The patterns are similar between winter and spring, demonstrating that northwestern wind control the north of 20°N and eastern wind control the south of 20°N. The eastern wind prevails at most of region in summer and fall, while south-western wind exists in the region between 30°N–40°N. Meanwhile, rainfall is the high in summer and low in winter and spring.

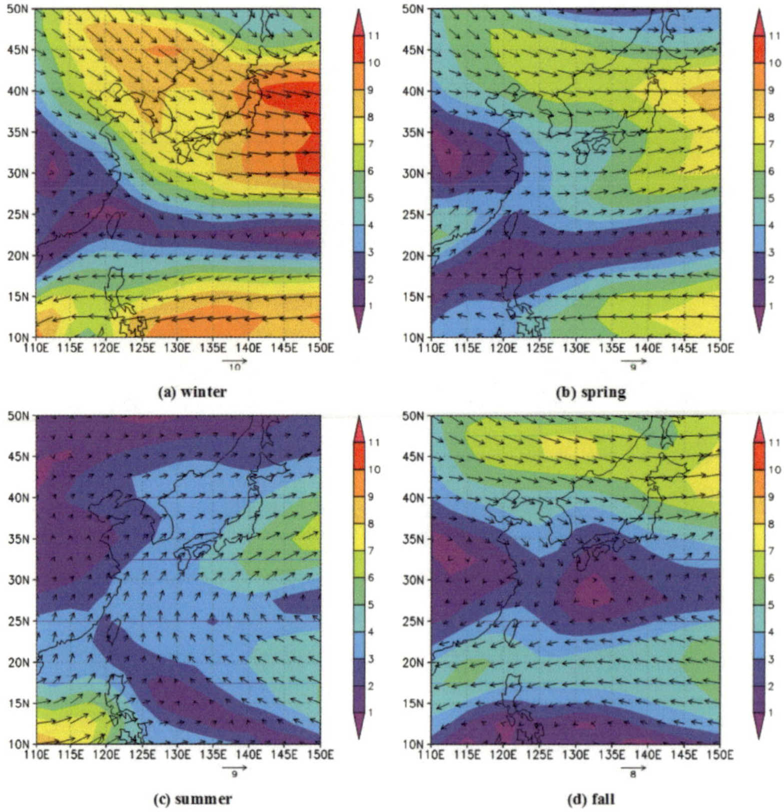

Figure 10. Seasonal distribution of wind field at 850 hp over the China Sea. (a) winters; (b) spring; (c) summer; (d) fall.

The aerosols are sent from the continental to ocean by the northwestern wind in winter. High wind speed causes that aerosol is difficult to bank up onto the China Sea, slowing down the values of the AOT. Meanwhile, high wind speeds create big particles of marine aerosol, leading to the smallest of the FMF. In spring, northwestern winds carry dust aerosols to the China Sea by and wind speed becomes weaker, so that aerosols build up onto the China Sea. The AOT reach maxima with a small of the FMF. In summer, east wind prevails, so continental aerosol can't be transported onto the sea. The rainfall is the large, bringing down the big particles of aerosols. So, the AOT is the smallest with the largest of the FMF. In fall, the west winds control the region of the north of 30° N, so the AOT is larger than summer. Rainfall is next below in summer, so the FMF is smaller than that in summer, but higher than that in other seasons. Therefore, meteorological conditions play an an important role on distribution of the AOT and FMF.

4. Aerosols monitored by satellites over the land

4.1. Data

Anhui province lies in the hinterland of Yangtze Delta in China between temperate zone and sub-tropical zone (**Figure 11**), one of the most important agricultural provinces covering 139,600 km^2. The agricultural activity and urbanization play important impacts on aerosols and air quality. For example, agricultural residues burning causes serious environment problem [32]. However, there is a little information related to aerosol characteristics with their influences of meteorological factors.

Figure 11. The geographical regions and the location of AERONET sites and meteorological stations.

One sample of SeaWiFS is selected and shown in **Figure 12**. It is a composite image of the satellite-received reflectance at the TOA on December 24, 2003. It covers a complicate structure of the ground reflectance including very turbid waters to oceanic waters, showing a mixture of reflectance from the Rayleigh, aerosols, ocean and land.

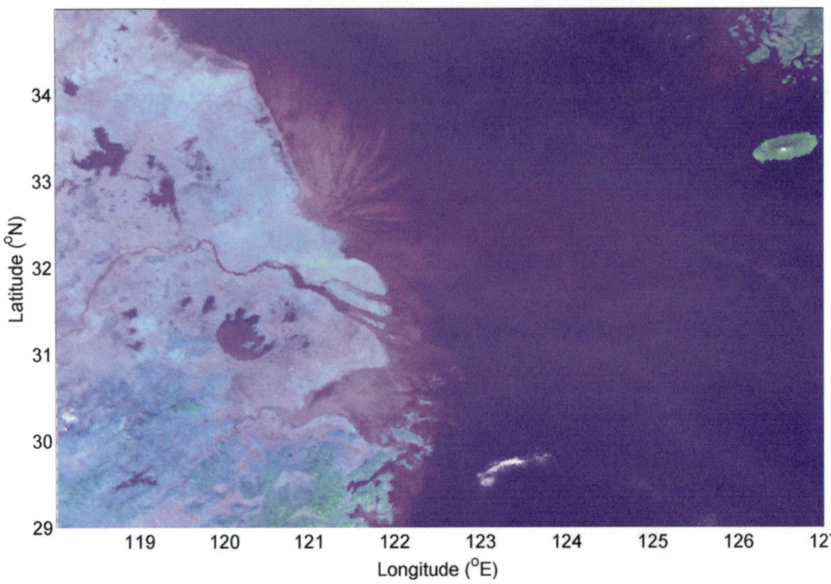

Figure 12. The image of the satellite-received reflectance at the TOA from SeaWiFS data on Dec. 24, 2003.

4.2. Some results

The UAC model uses the same computation method of the Rayleigh reflectance for the ocean and land, with the value is corrected by the elevation of the land surface. The aerosol reflectance obtained by the UAC model is shown in **Figure 13**, with significantly different structure from that of the ground reflectance. The image shows coherent spatial distributions extending from land into the coastal ocean, especially almost no difference in the aerosol distributions between lakes and nearby lands. The large spatial variations of the aerosol reflectance occur over the region around Shanghai and Lake Taihu. The spatial distribution of aerosols demonstrates that the UAC model has the advantage to estimate the aerosol reflectance without discontinuity between land and ocean.

4.3. Validating the retrieval aerosols over the land

There are more than 500 sites measured by AERONET around the world and four sites located in the region, which are EPA-NCU, Gosan_SNU, Gwangju_K-JIST and Taipei_CWB. The AOT data are used to obtain the aerosol scattering reflectance. They are matched with cloud free SeaWiFS data under the time difference of ±2 h at the same locations. The reflectance in Band 2 of the Gosan_SNU site is selected and shown in **Figure 14** according to the time of measurements.

http://dx.doi.org/10.5772/65284

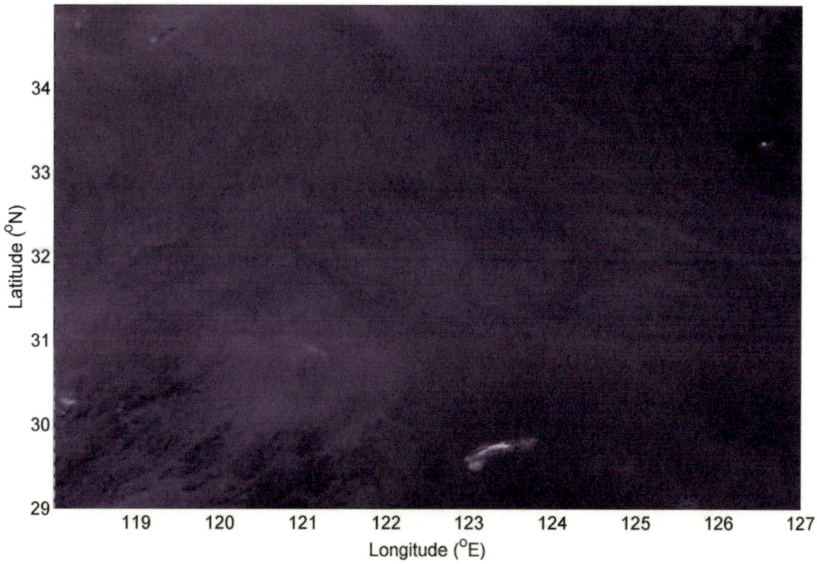

Figure 13. The aerosol reflectance obtained by the UAC model from SeaWiFS data on December. 24, 2003 shown with the RGB channels from Bands 6, 7 and 2.

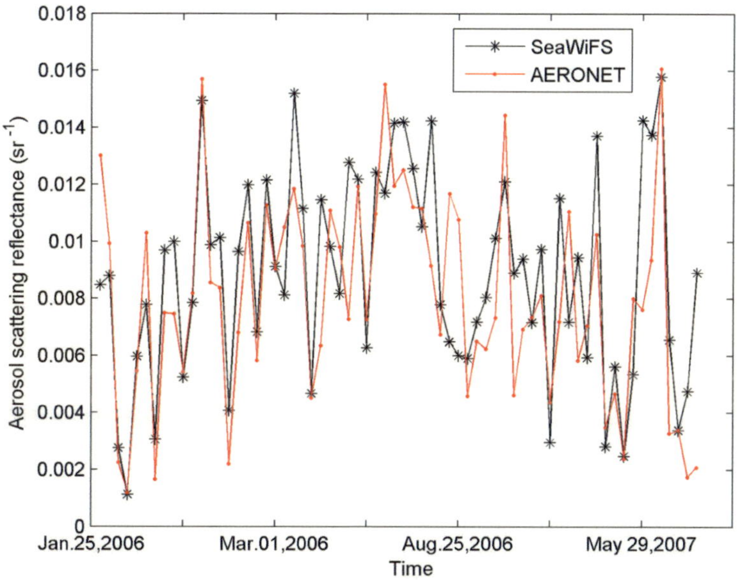

Figure 14. Comparison between the aerosol scattering reflectance from SeaWiFS data using the UAC model and that from the AERONET measurements in Band 2.

A total of 68 pairs of match-up data were established during 2006 and 2007 at the Gosan_SNU site. The aerosol scattering reflectance was obtained from SeaWiFS data using the UAC model and from the AOT data. The two reflectances in Band 2 vary consistently with each other, with the relative error of 21.9%. The reflectance in other four bands also displays similar results with the relative errors of 22.9, 23.8, 28.3 and 33.7%, respectively.

The aerosol scattering reflectance obtained from the AOT at other AERONET sites is also used to validate the accuracy of the reflectance from SeaWiFS data using the UAC model, shown in **Table 2**. Most of the relative errors are around 35%, with the mean values of 32.5, 33.3 and 33.6% corresponding to the sites of EPA-NCU, Gwangju_K-JIST and Taipei_CWB, respectively. The mean error of match-up pairs at the four sites is 31.4%. The standard deviations (STD) are almost the same, with only 2.8% difference. The two mean differences are the same, at 0.00085 sr^{-1}.

SeaWiFS (nm)	**412**	**443**	**510**	**670**	**865**
AERONET (nm)	380	440	500	675	865
Gosan_SNU (%)	22.88	21.94	23.84	28.29	33.74
EPA-NCU (%)	29.45	30.68	29.60	32.78	39.87
Gwangju_K-JIST (%)	28.30	28.51	30.62	36.14	43.16
Taipei CWB (%)	32.31	33.25	29.70	32.64	40.05

Table 2. The relative error of the SeaWiFS retrieval aerosol reflectance using the UAC model with respect to values calculated from the AERONET measured AOT data.

In order to take into account both spatial and temporal variations in aerosol distribution in Anhui, MODIS AOT at 10 km × 10 km resolution and AERONET AOT at 15-m intervals need to be co-located in space and time. In order to perform direct comparison and validation AOT at 550 nm, the values of AERONET AOT at 550 nm are retrieved by using Angstrom exponent calculated between 440 and 870 nm. The MODIS AOTs agree generally well with the AERONET AOTs. Most of dots overlay with the error range of $\Delta\tau = \pm 0.05 \pm 0.20\ \tau$. The linear fit slopes are less than 1, correlation coefficients are larger than 0.8, and RMS are less than 0.21 at four AERONET sites, which indicate that MODIS aerosol inversions have systematic biases in Anhui. The deviation from the unity of the slope of correlation plot represents systematic biases, whereas the intercept represents the errors due to surface reflectance assumptions [33]. The correlation coefficient results are all consistent.

4.4. Spatial and temporal variations of aerosols over the land

The spatial distribution and seasonal variations in AOT in Anhui averaged over the period 2001–2009 are presented in **Figure 15**. From the feature of aerosol spatial distribution, higher values of the AOT are observed in a large plain region at the north and middle of Anhui, while smaller values can been found over the mountainous region of southern Anhui centred over Huangshan and Dabieshan Mountain. We can find that the AOT is obviously influenced by topography with lower AOT over the mountainous region. Big particle aerosols are removed by gravitational settling with dominant small particle aerosols when terrain altitudes increase. The AOT is higher over the plain region, due to complex aerosol sources from dust, industry emissions, biomass burning and traffic sources.

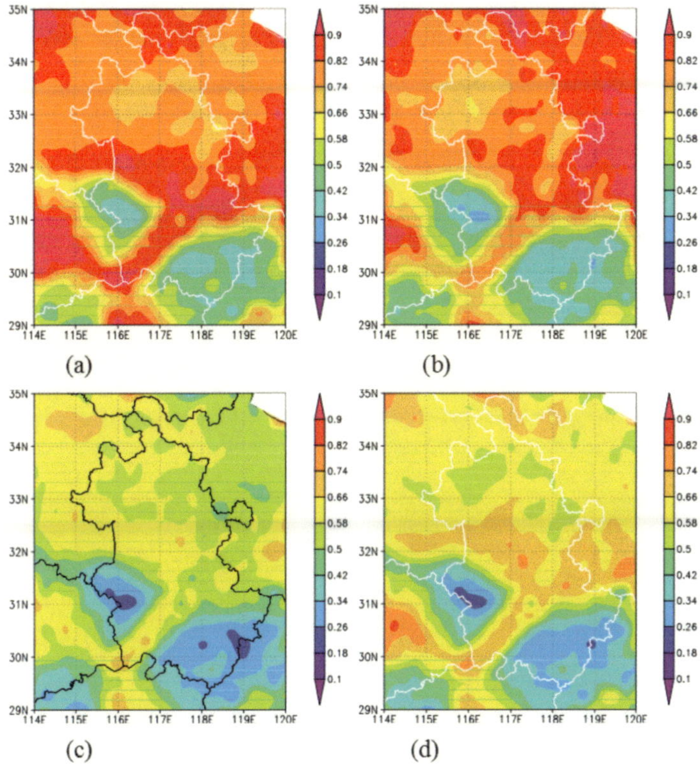

Figure 15. The seasonal mean of the AOT over Anhui province in China: (a) spring, (b) summer, (c) autumn and (d) winter.

The temporal variation of aerosol in Anhui can be well seen in **Figure 16**. It can be seen that there are obvious periodic variations in AOT, and the monthly variations are in agreement with the aforesaid result of seasonal changes. We find that the lowest AOT values are observed during the autumn with the highest during spring and summer. The high values of the aerosols are affected by dust originating from the northwestern China in spring. Despite that the amount of rainfall is maximum in summer, but the month of the highest AOT is June. The decrease in the AOT during autumn and winter is related with meteorological conditions.

4.5. The influence for aerosol by agricultural residues burning

The analysis shows that the month of the highest AOT is June. June is the beginning month of rainy season, in which AOT generally decreases by wet deposition. **Figure 17** shows a large increase (21%) in the AOT value from 0.77 in May to 0.93 in June and a prominent decrease

(32%) in AOT value from June (0.93) to July (0.63). The great change of AOT is likely from other sources of aerosols in June compared to in May and July. June is the time of agricultural residues burning after harvest periods in Anhui. Anhui province is a large agriculture area with about 30,000,000 tons of the grain output every year. The field burning is a general practice to clear the land for the next crop, controlling weeds and enriching soil nutrients. Thus, large amounts of aerosol particles are emitted by burning agricultural residues [34]. The fire pixels almost dominate the whole northern Anhui province and the fire pixel counts peak in June. Compared to the counts in other months, the fire pixel count is quite large in June and clearly matches with the high AOT loads during the burning month. As a result, the high AOT values observed in June in Anhui correspond to the burning agricultural.

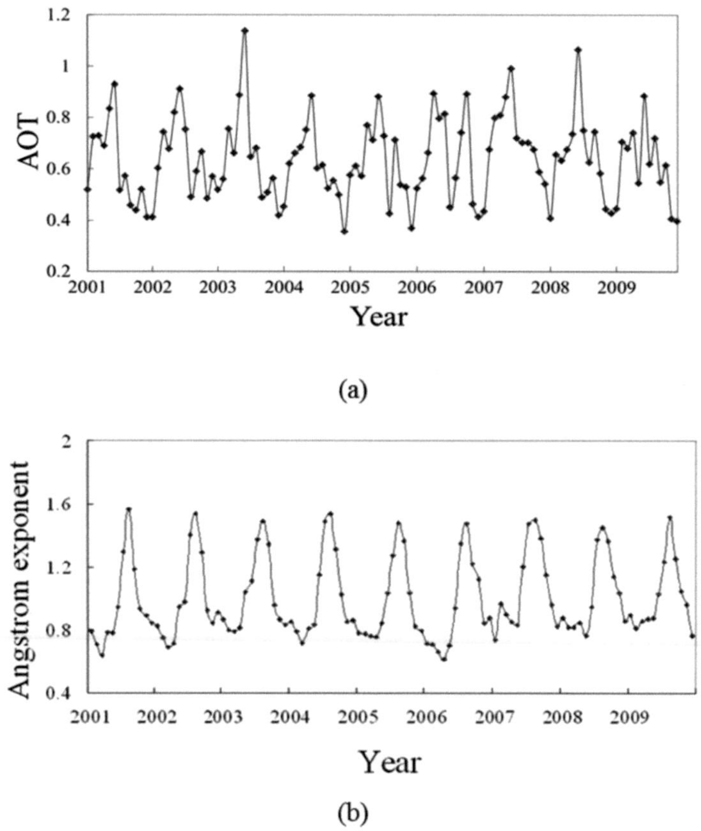

Figure 16. Monthly mean AOT and Angstrom exponent over Anhui province in China: (a) AOT and (b) Angstrom exponent.

4.6. Back trajectory analysis

The 3-day back trajectory analysis on 850 hPa was calculated to examine the aerosol sources of different AOT by using the hybrid single-particle Lagrangian integrated trajectory

http://dx.doi.org/10.5772/65284

(HYSPLIT) model. In order to obtain the influence for different aerosol particle size ranges by trajectories, two categories of back trajectories are defined based on values of the Angstrom exponent: coarse mode aerosol (CMA) with Angstrom exponent less than 1 and fine mode aerosol (FMA) with Angstrom exponent over 1, in order to obtain the influence for (**Figure 18**). A similar analysis for AOT degrees and the Angstrom exponent levels has been conducted.

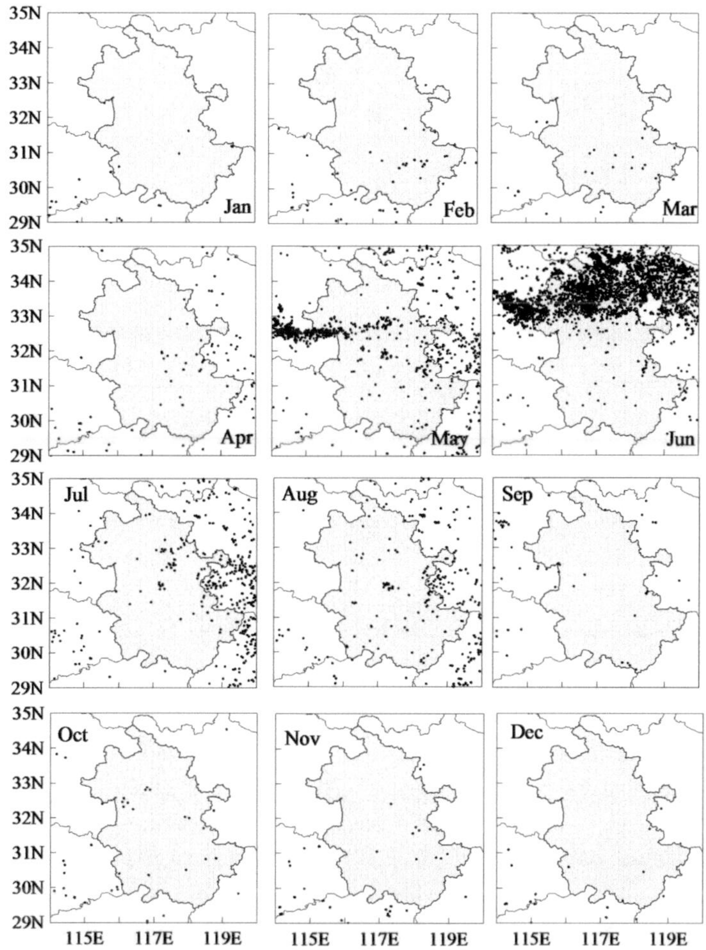

Figure 17. Monthly variation of fire pixels in 2007 in Anhui province from MODIS global monthly fire location product (MCD14ML). Fire pixels are shown in black dots.

For CMA category (**Figure 18a**), 8 clusters were obtained. The northwest transport pathway (Clusters 1, 2, 5–8) contributed 70% of all the trajectories, appearing with coarse mode aerosol. Long northwest clusters stand out with abundance of dusts, and dusts are carried by the northwest trajectories from dust sources to Fuyang. Therefore, aerosol particle sizes are mainly influenced by the northwest flows.

For FMA category (**Figure 18b**), six clusters were observed. Clusters 2 (13%) and 3 (8%) represent the northwest flows, while Cluster 1 represents the flow of the southwest sector (17%). Clusters 4 (23%), 5 (19%) and 6 (20%) have westerly, easterly and north easterly origins, respectively. The most notable difference of trajectories between CMA and FMA is the distance of transport flows. The long northwest trajectories (70%) are dominant for CMA, while the short transport flows (62%) are the most for FMA. This result is attributed to different aerosol types under different air masses. Therefore, anthropogenic aerosols mainly lead to fine mode, and natural aerosols dominate the long northwest trajectories.

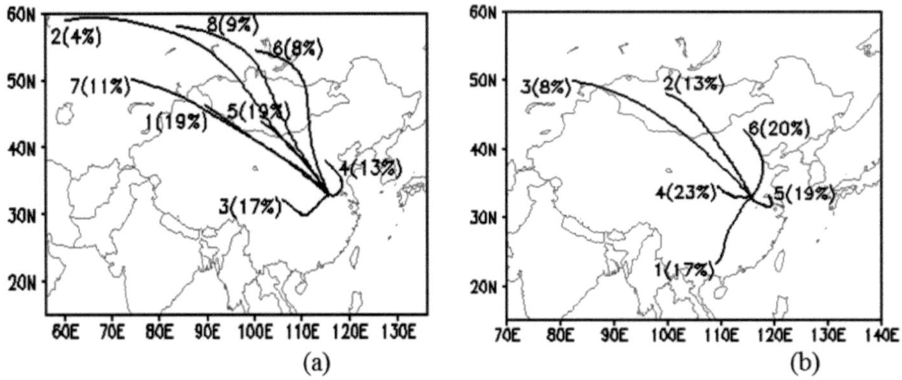

Figure 18. Mean HYSPLIT 72-h backward trajectories of 1500 m altitude at 02:30 GMT by cluster analysis during 2001–2009 at Fuyang station. (a) CMA, Angstrom exponent <1 and (b) FMA, Angstrom exponent > 1.

5. Aerosol vertical structures monitored by satellite lidar

5.1. Introduction

The CALIPSO combines an active lidar instrument with passive infrared and visible imagers to probe the vertical structure and properties of thin clouds and aerosols over the globe [35], to study the climate impact of clouds and aerosols in the atmosphere [36]. The Cloud-Aerosol Lidar with Orthogonal Polarization (CALIOP) is the primary instrument on the CALIPSO satellite, which is the first polarization lidar in orbit and the first satellite lidar [37]. It provides vertically resolved measurements of aerosol distribution with aerosol extinction coefficients [38] and observes aerosols over bright surfaces and beneath thin clouds as well as in clear sky conditions [39].

The CALIOP lidar consists of a laser and a receiver with the stability of the transmitter-to-receiver alignment [40]. The lasers produce simultaneous pulses at 532–1064 nm at a pulse repetition rate of 20.16 Hz, with a pulse length of about 20 ns. Polarization out-coupling provides a highly polarized beam with a beam diameter of 70 m at the Earth's surface [36]. A polarization beam splitter separates the 532 nm parallel and perpendicular returns. The receiver consists of the 1-m telescope and the photomultiplier tubes (PMTs).

http://dx.doi.org/10.5772/65284

The CALIOP acquires vertical profiles of elastic backscatter at two wavelengths from a near nadir-viewing geometry during both day and night phases of the orbit, to derive the accurate aerosol and cloud heights and extinction coefficient profiles. It measures the profiles of linear depolarization at 532 nm, to discriminate ice clouds from water clouds and to identify non-spherical aerosol particles [41]. The aerosol particle sizes can be obtained from the ratios of the signals obtained at the two wavelengths [42].

5.2. The algorithms to retrieve the aerosol vertical structures

The fundamental algorithms of the CALIPSO are profile processes, to locate all layer boundaries and to identify each layer as being either cloud or aerosol. There are three types of algorithms related with the layer detection, the feature classification and the optical properties analysis. The data products include layer heights (e.g. feature top and base altitudes), layer identification (i.e. clouds versus aerosols, ice clouds versus water clouds), layer structures of clouds, aerosol backscatter and extinction coefficients.

The first step is to identify the layer types of clouds and aerosols. Thus the layer detection algorithm is used to identify regions of enhanced scattering, and to record simple characteristics of these atmospheric features, and to detect the vertical location of several different classes of geophysical objects. Specifically, the lidar returns are used to obtain information on the base and top altitudes of clouds and aerosol layers [43]. The layer finding algorithm is based primarily on separating the genuine features from the pseudo-features to find layers repeatedly on several passes through the data. The distribution of the roughly separated aerosols with clouds is shown in **Figure 19**. The regions with white are recognized as clouds and the regions with yellow are taken as aerosols. The intensified backscattering values of clouds are centred at 1.5 km and those of aerosol layers are centred at 0.5 km.

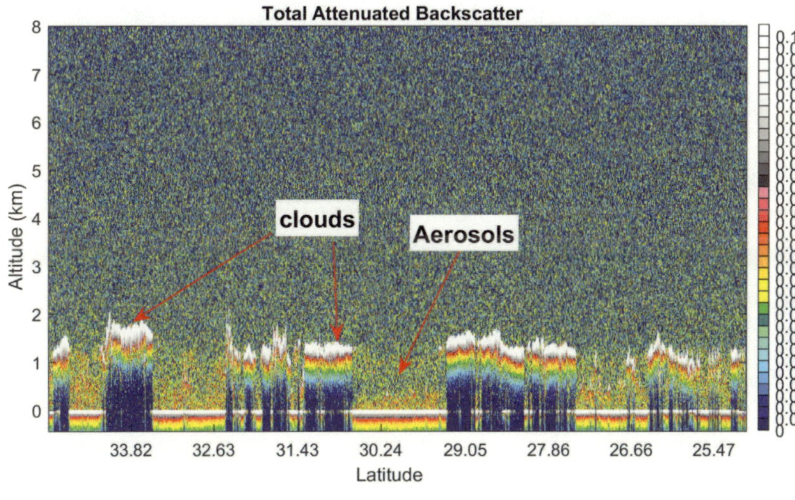

Figure 19. Aerosols and clouds can be roughly separated from the satellite data on Aug. 18, 2015.

5.3. Scene classification algorithm

The second processing step is to classify the layers and identify the types. The scene classification algorithm (SCA) is used to determine the types of the features (tropospheric or stratospheric) by checking the base altitude of the feature. The tropopause altitude is derived from ancillary data obtained from the global modelling and assimilation office and used to classify the tropospheric feature depending on whether the feature base is lower than this altitude, while the stratospheric feature is classified by that higher than this altitude. The further classification algorithms are conducted to sub-type the feature for the tropospheric feature.

The SCA is used to determine the aerosol layer or cloud layer, primarily based on the scattering strength and the spectral dependence of the lidar backscattering, which are primarily obtained from the mean value of the attenuated backscatter coefficient and the ratio of the mean attenuated backscatter coefficients (the attenuated colour ratio) at 1064–532 nm (**Figure 20**). If the layer is classified as cloud, the SCA will then determine whether it is an ice cloud or water cloud using the backscatter intensity and the depolarization ratio profiles together with ancillary information of the layer height and temperature. The SCA will also use a combination of observed parameters and a priori information to select an appropriate extinction-to-backscatter ratio and multiple scattering function required for subsequent process of the extinction and optical depth retrieve.

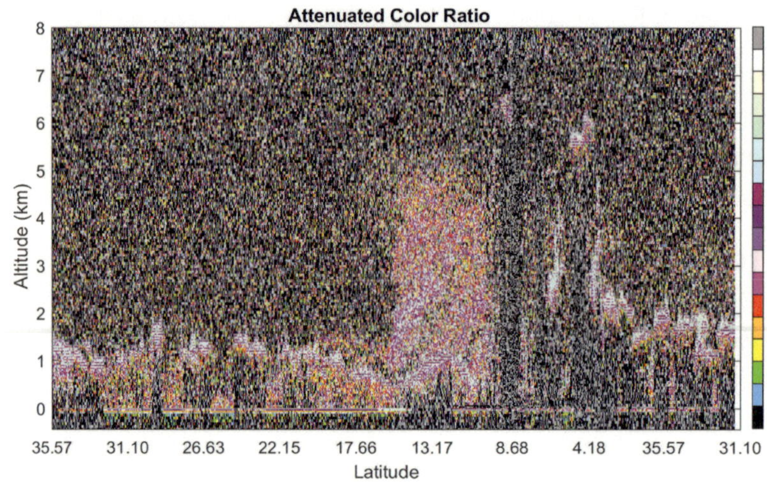

Figure 20. The attenuated colour ratio at 1064–532 nm for determining the layer type.

5.4. Retrieved optical properties analysis

The last processing step is to retrieve the extinction coefficient and optical depth from the calibrated, range-corrected lidar signal. Some widely used algorithms have been developed, based on the Fernald method [44], the Klett method [45] and the so-called linear iterative method [46]. The Fernald and Klett methods offer analytic solutions with closed form, while the linear iterative technique is a simple numerical solution. Both Fernald and Klett methods are developed from the single scattering analyses to the multiple scattering analyses using a correction factor to the range-resolved extinction coefficients. The CALIPSO algorithms account for multiple scattering by applying a correction factor derived from the phase functions of aerosol models [47]. These methods are used to retrieve the extinction and backscatter profiles of cloud and aerosol layers.

After the layer profiles identified by layer detection algorithm and classified by the SCA, the hybrid extinction retrieval algorithm (HERA) performs extinction retrievals on regions of profile data [48]. HERA incorporates a sophisticated retrieval engine that extracts profiles of particulate backscatter and extinction coefficients from the profiles of attenuated backscatter coefficients identified as layers. One example of particulate backscatter and extinction at 532–1064 nm is shown in **Figure 21**. Following the retrieval of extinction profiles, other parameters can be computed by HERA, including optical depth, particle colour ratio and particle depolarization ratio [49].

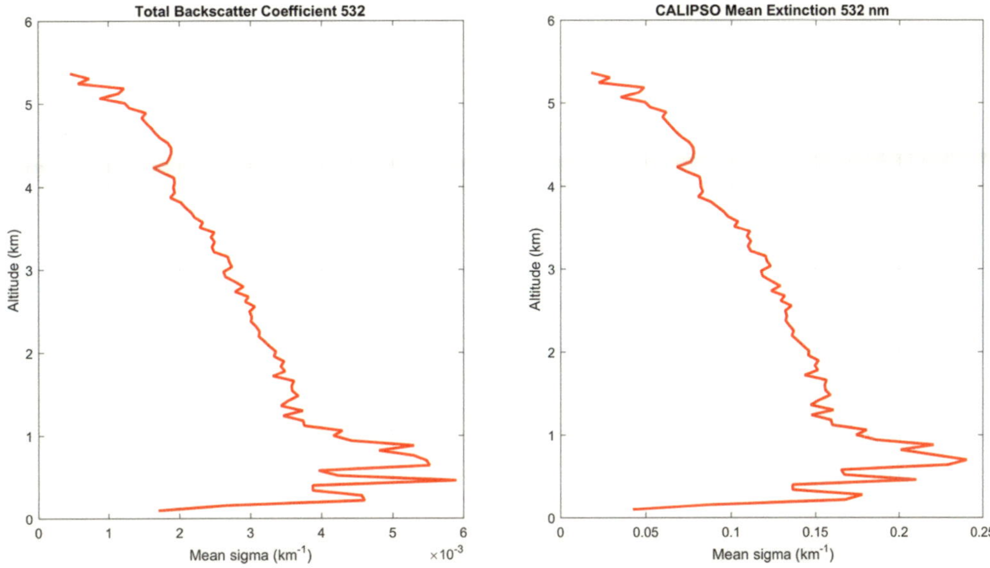

Figure 21. The left panel means profile of mean extinction at 532 nm; the right panel means profile of attenuation-corrected backscatter coefficient at 532 nm.

Column atmospheric optical depth can be estimated from the measured two-way transmittance at the ocean surface using ocean surface backscatter and collocated wind speed. Once the collocated wind speed is determined, theoretical ocean surface lidar backscatter can be derived from the wave slope variance—wind speed relation. The column optical depth is then derived from the ratio of the CALIPSO attenuated backscatter measurement and theoretical backscatter [50]. The cloud and aerosol optical depth derived from this approach is a direct measurement, without assuming aerosol and cloud physical properties. An example of aerosol optical depth distribution retrieved from CALIPSO data can be seen in **Figure 22**.

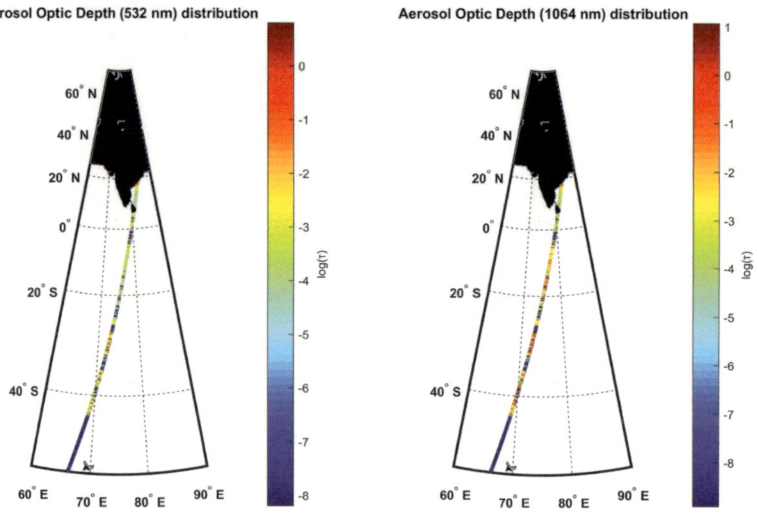

Figure 22. An example of AOD distribution retrieved from CALIPSO data with the AOD at 532 nm (the left panel) and 1064 nm (the right panel).

6. Conclusions

The ground reflectance is significantly different over land and ocean, leading to the need for different approaches of the atmospheric correction in order to optimize the results. However, this difference can be expressed by the spectra of the LUT of in situ measurements in the UAC model. Based on the LUT of the ground reflectance, the aerosol reflectance can be obtained by the best nonlinear least square fit function based on the Angstrom law. The reflectance is then used to determine the epsilon spectra that are used to select the two closest aerosol models instead of one epsilon value from the NIR band. The results show that this approach is more robust to overcome the problem caused by small variations in the aerosol reflectance in the two NIR bands. Meanwhile, this approach can also reduce some abnormal epsilon errors caused by factors including the data quality of satellite reflectance and the mismatch between the actual ground reflectance and the LUT. The use of the LUT provides a unified approach for estimating the aerosol reflectance and the ground reflectance over land and ocean. The performance of the UAC model is evaluated using a SeaWiFS image. The results show that the model can completely separate the aerosol scattering reflectance from the radiance at TOA. The relative error is 22.1% when it is validated by the in situ measured AOT data using the MICROTOP instruments, and the error is 31.4% using the four sites of AERONET measured AOT data over the ECS in 2006 and 2007. The UAC model can provide the aerosol products from satellite remote sensing data over land and ocean.

MODIS AOTs at 12 stations around China using the AERONET observations have been evaluated with the result that both agree generally well over ocean and land. The seasonal changes and spatial distributions of aerosol optical properties have also been observed with a notable

http://dx.doi.org/10.5772/65284

temporal variation over the China Sea. The AOT reaches maxima in spring and winter and minima in summer and fall, while the FMF reaches maxima in summer and fall and minima in spring and winter. Both AOT and FMF have an obvious spatial distribution over the China Sea. At latitude direction, the AOT appears maxima between 30 and 40°N and the FMF increases from south to north. At longitude direction, both AOT and FMF decrease with longitudinal increasing. Over the land, the lowest AOT values are observed during the autumn and the highest during the spring. The highest AOT, appearing in June, is possibly related to burning of agricultural residues. The Angstrom exponent appears maxima in summer and minima in winter. The higher AOT and lower Angstrom exponent are observed in a large plain region at the north and middle, while smaller AOT and higher Angstrom exponent can be found over the mountainous region, which is obviously influenced by topography. Meteorological conditions obviously affect the aerosol optical properties. Continent aerosol is carried to the China Sea by wind. The influences of different air masses are great disparity for aerosol optical properties over the land.

Acknowledgements

This study is supported by the National Key Research and Development Program of China (2016YFC1400901), the National Science Foundation of China (41476156, 41321004), and the Public Science and Technology Research Funds Projects of Ocean (201005030), the Anhui research project in the Public Interest (1604f0804003), and the China Special Fund for Meteorological Research in the Public Interest (GYHY201406039). We thank our colleagues from the Second Institute of Oceanography SOA, who took part in the cruises, for their hard work on in situ measurements. We also want to thank the experts who provided us their in-situ measurements.

Author details

Zhihua Mao[1,2*], Xueliang Deng[3], Peng Chen[1], Bangyi Tao[1], Guanying Yang[3], Yanfeng Huo[3] and Qiankun Zhu[1]

*Address all correspondence to: mao@sio.org.cn

1 State Key Laboratory of Satellite Ocean Environment Dynamics, Second Institute of Oceanography, State Oceanic Administration, Bochubeilu, Hangzhou, China

2 Nanjing University, Nanjing, China

3 Key Laboratory of Atmospheric Science and Satellite Remote Sensing, Anhui Institute of Meteorology, Hefei, China

References

[1] Griggs D J, Noguer M. Climate change 2001: The scientific basis. Contribution of Working Group I to the Third Assessment Report of the Intergovernmental Panel on Climate Change[J]. Weather, 2002, 57(8):267–269.

[2] Penner J E, Charlson R J, Schwartz S E, et al. Quantifying and minimizing uncertainty of climate forcing by anthropogenic aerosols. Bulletin of the American Meteorological Society, 1994, 75(3): 375–400.

[3] Satheesh S K, Moorthy K K. Radiative effects of natural aerosols: A review. Atmospheric Environment, 2005, 39(11): 2089–2110.

[4] Ramanathan V, Crutzen P J, Kiehl J T, et al. Aerosols, climate, and the hydrological cycle. Science, 2001, 294(5549): 2119–2124.

[5] Pope III C A, Burnett R T, Thun M J, et al. Lung cancer, cardiopulmonary mortality, and long-term exposure to fine particulate air pollution. JAMA, 2002, 287(9): 1132–1141.

[6] Winker D, Vaughan M, Hunt B. The CALIPSO mission and initial results from CALIOP[C]//Asia-Pacific Remote Sensing Symposium. International Society for Optics and Photonics, 2006: 640902–640902-8.

[7] Kaufman Y J, Tanré D, Boucher O. A satellite view of aerosols in the climate system. Nature, 2002, 419(6903): 215–223.

[8] Gordon H R. Removal of atmospheric effects from satellite imagery of the oceans. Applied Optics, 1978, 17(10): 1631–1636.

[9] Gordon H R, Brown J W, Evans R H. Exact Rayleigh scattering calculations for use with the Nimbus-7 Coastal Zone Color Scanner. Applied Optics, 1988, 27(5): 862–871.

[10] Ding K, Gordon H R. Analysis of the influence of O(2) A-band absorption on atmospheric correction of ocean-color imagery. Applied Optics, 1995, 34(12): 2068–2080.

[11] Wang M, Wei S. Estimation of ocean contribution at the MODIS near-infrared wavelengths along the east coast of the US: Two case studies[J]. Geophysical Research Letters, 2005, 32(13):370–370.

[12] Ruddick K G, Ovidio F, Rijkeboer M. Atmospheric correction of SeaWiFS imagery for turbid coastal and inland waters. Applied Optics, 2000, 39(6): 897–912.

[13] Shanmugam P, Ahn Y H. New atmospheric correction technique to retrieve the ocean colour from SeaWiFS imagery in complex coastal waters. Journal of Optics A: Pure and Applied Optics, 2007, 9(5): 511.

[14] Siegel D A, Wang M, Maritorena S, et al. Atmospheric correction of satellite ocean color imagery: The black pixel assumption. Applied Optics, 2000, 39(21): 3582–3591.

http://dx.doi.org/10.5772/65284

[15] Wang M. Remote sensing of the ocean contributions from ultraviolet to near-infrared using the shortwave infrared bands: Simulations. Applied Optics, 2007, 46(9): 1535–1547.

[16] He X, Bai Y, Pan D, et al. Using geostationary satellite ocean color data to map the diurnal dynamics of suspended particulate matter in coastal waters. Remote Sensing of Environment, 2013, 133: 225–239.

[17] Kaufman Y J, Sendra C. Algorithm for automatic atmospheric corrections to visible and near-IR satellite imagery. International Journal of Remote Sensing, 1988, 9(8): 1357–1381.

[18] Hall F G, Strebel D E, Nickeson J E, et al. Radiometric rectification: Toward a common radiometric response among multidate, multisensor images. Remote Sensing of Environment, 1991, 35(1): 11–27.

[19] Richter R. A spatially adaptive fast atmospheric correction algorithm. International Journal of Remote Sensing, 1996, 17(6): 1201–1214.

[20] Gao B C, Montes M J, Davis C O, et al. Atmospheric correction algorithms for hyper-spectral remote sensing data of land and ocean. Remote Sensing of Environment, 2009, 113(9): S17–S24.

[21] Mao Z, Chen J, Hao Z, et al. A new approach to estimate the aerosol scattering ratios for the atmospheric correction of satellite remote sensing data in coastal regions. Remote Sensing of Environment, 2013, 132: 186–194.

[22] Mao Z, Pan D, Hao Z, et al. A potentially universal algorithm for estimating aerosol scattering reflectance from satellite remote sensing data. Remote Sensing of Environment, 2014, 142: 131–140.

[23] Mao Z, Pan D, He X, et al. A unified algorithm for the atmospheric correction of satellite remote sensing data over land and ocean. Remote Sense, 2016, 8: 536.

[24] King M D, Kaufman Y J, Tanré D, et al. Remote sensing of tropospheric aerosols from space: Past, present, and future. Bulletin of the American Meteorological Society, 1999, 80(11): 2229–2259.

[25] Winker D M, Pelon J, Coakley Jr J A, et al. The CALIPSO mission: A global 3D view of aerosols and clouds. Bulletin of the American Meteorological Society, 2010, 91(9): 1211.

[26] Wang M. Aerosol polarization effects on atmospheric correction and aerosol retrievals in ocean color remote sensing. Applied Optics, 2006, 45(35): 8951–8963.

[27] Gordon H R, Castaño D J. Aerosol analysis with the coastal zone color scanner: A simple method for including multiple scattering effects. Applied Optics, 1989, 28(7): 1320–1326.

[28] Mao Z, Chen J, Pan D, et al. A regional remote sensing algorithm for total suspended matter in the East China Sea. Remote Sensing of Environment, 2012, 124: 819–831.

[29] Holben B N, Eck T F, Slutsker I, et al. AERONET—A federated instrument network and data archive for aerosol characterization. Remote Sensing of Environment, 1998, 66(1): 1–16.

[30] Smirnov A, Holben B N, Eck T F, et al. Cloud-screening and quality control algorithms for the AERONET database. Remote Sensing of Environment, 2000, 73(3): 337–349.

[31] Mélin F, Vantrepotte V. How optically diverse is the coastal ocean. Remote Sensing of Environment, 2015, 160: 235–251.

[32] He L M, Wang W J, Wang Q, et al. Evaluation of the agricultural residues burning reduction in China using MODIS fire product. Environmental Monitoring in China, 2007, 23(1): 42–50.

[33] Remer L A, Kaufman Y J, Tanré D, et al. The MODIS aerosol algorithm, products, and validation. Journal of the Atmospheric Sciences, 2005, 62(4): 947–973.

[34] Zhang H, Ye X, Cheng T, et al. A laboratory study of agricultural crop residue combustion in China: Emission factors and emission inventory. Atmospheric Environment, 2008, 42(36): 8432–8441.

[35] Sassen K, Wang Z, Liu D. Global distribution of cirrus clouds from CloudSat/Cloud-Aerosol Lidar and Infrared Pathfinder Satellite Observations (CALIPSO) measurements[J]. Journal of Geophysical Research Atmospheres, 2008, 113(8):347–348.

[36] Lu X, Hu Y, Trepte C, et al. Ocean subsurface studies with the CALIPSO space borne lidar. Journal of Geophysical Research: Oceans, 2014, 119(7): 4305–4317.

[37] Winker D M, Hostetler C A, Vaughan M A, et al. CALIOP algorithm theoretical basis document, part 1: CALIOP instrument, and algorithms overview. Release, 2006, 2: 29.

[38] Liu D, Wang Z, Liu Z, et al. A height resolved global view of dust aerosols from the first year CALIPSO lidar measurements[J]. Journal of Geophysical Research Atmospheres, 2008, 113(D16):280–288.

[39] Geng F, Liu Q, Chen Y, et al. Preliminary study of vertical distribution of aerosols during dry haze periods around Shanghai based on CALIPSO. Procedia Earth & Planetary Science, 2011, 2(1): 217–222.

[40] Winker D M, Vaughan M A, Omar A, et al. Overview of the CALIPSO mission and CALIOP data processing algorithms. Journal of Atmospheric and Oceanic Technology, 2009, 26(11): 2310–2323.

[41] Winker D M, Hunt W H, Mcgill M J. Initial performance assessment of CALIOP[J]. Geophysical Research Letters, 2007, 34(19):228–262.

[42] Lindberg J D, Loveland R B, Duncan L D, et al. Vertical Profiles of Extinction and Particle Size Distribution Measurements Made in European Wintertime Fog and Haze[J]. ASL-TR-0151, 1984,1: 88002.

[43] Mark A. Vaughan, David M. Winker, Kathleen A. Powell, et al. Fully automated analysis of space-based lidar data: an overview of the CALIPSO retrieval algorithms and data products[J]. Proceedings of SPIE - The International Society for Optical Engineering, 2004, 5575:16–30.

[44] Fernald F G. Analysis of atmospheric lidar observations: Some comments. Applied Optics, 1984, 23(5): 652–653.

[45] Klett J D. Lidar inversion with variable backscatter/extinction ratios. Applied Optics, 1985, 24(11): 1638–1643.

[46] Elterman L. Aerosol measurements in the troposphere and stratosphere. Applied Optics, 1966, 5(11): 1769–1776.

[47] Omar A H, Winker D M, Won J G. Aerosol models for the CALIPSO lidar inversion algorithms[C]//Remote Sensing. International Society for Optics and Photonics, 2004, 1: 153–164.

[48] Vaughan M A, Winker D M, Hostetler C A. SIBYL: a selective iterated boundary location algorithm for finding cloud and aerosol layers in CALIPSO lidardata[J]. Lidar Remote Sensing in Atmospheric and Earth Sciences, 2002, 1: 791–794.

[49] Young S A, Winker D M, Vaughan M A, et al. CALIOP algorithm theoretical basis document, part 4: Extinction retrieval algorithms[J]. NASA Langley Research Center, Tech. Rep, Hampton, Virginia, 2008.

[50] Hu Y. Ocean, Land and Meteorology Studies Using Space-Based Lidar Measurements[C]. Proceedings of the 5th WSEAS International Conference on REMOTE SENSING,2009,1:47–50.

Chapter 6

The Image-Based Integrated Method for Determining and Mapping Aerosol Optical Thickness

Kyriacos Themistocleous

Additional information is available at the end of the chapter

http://dx.doi.org/10.5772/65279

Abstract

This chapter focuses on the development of an image-based integrated method for determining and mapping aerosol optical thickness (AOT). Using the radiative transfer (RT) equation, a methodology is developed using nonvariant targets with the revised darkest pixel method to improve the accuracy to calculate AOT over urban areas. This method can be applied to create thematic maps using GIS that present AOT values in a given area. The methodology is applied to Landsat TM/ETM+ satellite images of Limassol, Cyprus over a period of time to visualize and assess the AOT levels over the urban area. An accuracy assessment of the method shows a strong correlation between the *in-situ* AOT values from the sunphotometers and the AOT values derived from the image-based integrated method.

Keywords: AOT, GIS, atmospheric correction, image-based mapping, Landsat, Kriging

1. Introduction

Aerosol optical thickness (AOT) is defined as the degree to which aerosols prevent the transmission of light through absorption or scattering of light. AOT can be retrieved by using radiometers on board satellites [1–3], ground-based sunphotometers, and satellite sensors, such as the moderate resolution imaging spectroradiometer(MODIS), which provide AOT measurements on a daily and monthly basis [4, 5]. However, the spatial resolution of the MODIS AOT data products is 10 × 10 km, which does not permit the identification of specific AOT distribution over urban areas or complex terrains [5–7]. Due to the variability of aerosols, atmospheric aerosol monitoring is difficult. Significant efforts to improve aerosol characterizations have included using *in-situ* measurements, ground-based remote sensing, and satellite observations [1–3]. AOT derived from satellite images often requires further validation [8]. The accuracy of satellite-derived AOT is frequently assessed by comparing satellite-based AOT with **AE**rosol **RO**botic **NET**work (AERONET, a program established by **NASA**) or field-based sun photometers [9, 10].

Research indicates that the determined AOT from satellite image data can be used as a tool to assess air pollution as well as identify the sources of local emissions [2, 7, 11–24]. AOT values can be used as a way of measuring air quality; determining AOT in large-scale pollution areas provides a synoptic, cost-effective means to further assess the air quality in such areas [16, 18–24]. Indeed, the AOT values derived from the atmospheric path radiance can be utilized to assess and monitor air quality and atmospheric pollution [8]. The image-based integrated method presented in this chapter can accurately calculate AOT values retrieved from satellite imagery by using radiative transfer (RT) equations and GIS. The method can be used to visually display AOT levels using thematic maps in order to identify concentrations of AOT over an urban area [11]. The image-based method can also be used with archived satellite images, thereby providing detailed information regarding spatial aerosol concentration overtime [11].

AOT is directly related to the atmospheric aerosol load, which is the main variable describing the effects of aerosols on radiative transfer in the Earth's atmosphere. According to Guanter et al. [25], modeling atmospheric constituents and surface reflectance involves modeling the radiative transfer across the atmosphere. The key parameter for assessing atmospheric pollution is the aerosol optical thickness, which is also the most important unknown of every atmospheric correction algorithm for solving the radiative transfer equation and removing atmospheric effects from remotely sensed satellite images [8, 26–29]. Several researchers [5, 18, 26, 28, 29–37] found that using the radiative transfer and atmospheric modeling in conjunction with field measurements of aerosol optical thickness can yield more accurate atmospheric corrections instead of using simple image-based techniques.

The image-based integrated method discussed in this chapter combines the RT equation, satellite imagery, the modified darkest pixel (DP) method of atmospheric correction, and GIS to derive AOT measures. An example including 11 Landsat satellite images with *in-situ* measurements over a specific period of time is used to assess the AOT values based on the 30 × 30 m spatial resolution of Landsat over the city of Limassol, Cyprus and create thematic maps to display the AOT levels.

2. Methodology

The innovation of this methodology is the retrieval of AOT by using an image-based algorithm based on the radiative transfer equation to derive AOT using the reflectance values from the darkest pixel method of atmospheric correction. In the image-based integrated method, the RT equation retrieves AOT values from the satellite images with as much precision as possible. Also, the integrated use of field spectroscopy, GIS, and remote sensing for AOT retrieval is an important contribution to the field of remote sensing.

http://dx.doi.org/10.5772/65279

The first step of the methodology is the preprocessing of the satellite images, including geometric and radiometric correction. Also, the satellite images need to be atmospherically corrected. The darkest pixel atmospheric correction method can be applied, as it provides reasonable atmospheric correction, especially in cloud-free skies [28, 38, 39]. If there is a known nonvariant target or overpass *in-situ* measurements available in the selected site where the true ground reflectance values of the darkest target are known, then the modified DP method can be applied to result in more accurate reflectance values to the corrected satellite image [40]. The images are then processed using the RT equations in order to retrieve the AOT levels. Specific parameters from the satellite images, sun photometer measurements, and the reflectance values from the atmospherically corrected images are incorporated. In this methodology, the AOT values are calculated at the 500 nm wavelength, so Landsat band 1 will be used to derive AOT levels through the algorithm. Also, GIS analysis is done to produce thematic maps showing the AOT levels in the area of interest. In this chapter, an accuracy assessment is performed in the example of Limassol, Cyprus, where the AOT values retrieved from the GIS analysis are compared with the *in-situ* sunphotometer measurements from the sunphotometers. **Figure 1** shows the methodology for the AOT retrieval.

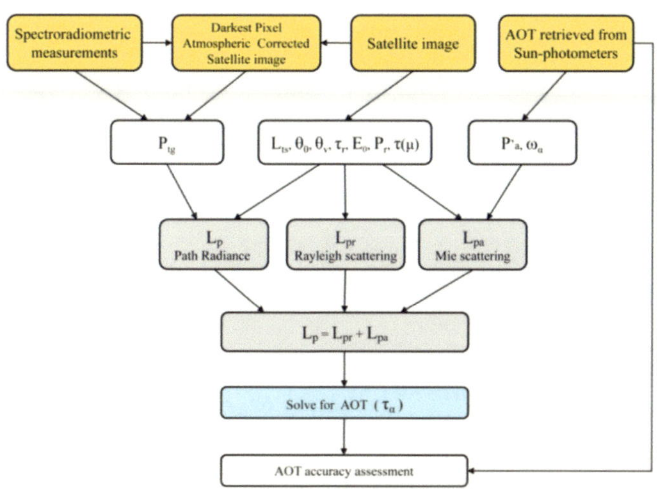

Figure 1. Methodology for AOT retrieval.

2.1. Detailed description of the image-based integrated method for determining AOT

According to Hadjimitsis and Clayton [40], RT equations can be used to retrieve AOT values from the path radiance. Therefore, the algorithm used in the image-based integrated method is formulated to solve for AOT, which is included in the equation for path radiance (L_p) and Mie scattering [11]. To calculate AOT, Eq. (1) is used.

Path radiance is the sum of the atmospheric path radiance due to Rayleigh (L_{pr}) and Mie scattering (L_{pa}) [11, 29, 40–43]:

$$L_p = L_{\text{pr}} + L_{\text{pa}} \tag{1}$$

where L_p is the path radiance (integrated band radiance measured in W/m^{2}sr), L_{pr} is the atmospheric path radiance due to Rayleigh scattering in W/m^{2}sr [44], and L_{pa} is the atmospheric path radiance due to Mie scattering in W/m^{2}sr [44].

2.1.1. Calculating path radiance

The path radiance "L_p" measured in Wm^{-2} μm^{-1} sr^{-1}, is calculated according to Eq. (2) [42, 45]:

$$L_p = L_{\text{ts}} - \frac{\tau(\mu)\uparrow \rho_{\text{tg}} E_G}{\pi}, \tag{2}$$

where L_{ts} is the measured at-satellite radiance, $\tau(\mu)\uparrow$ is the direct (upward) target sensor atmospheric transmittance, ρ_{tg} is the ground target reflectance, and E_G is the global irradiance reaching ground.

Eq. (3) can be used to calculate the at-satellite radiance (L_{ts}) for Landsat satellite images that contain gain and calibration offset values in the satellite image's header file, as described in the Landsat-7 Handbook [46]:

$$L_{\text{ts}} = \text{gain} \times \text{DN} + \text{offset} \tag{3}$$

The "E_G" is calculated according to Eq. (4) [45]:

$$E_G = E_{0\lambda}\mu exp\left[-\left(\frac{1}{2\tau_r} + \frac{1}{6\tau_a}\right)/\mu\right] \tag{4}$$

where $E_{0\lambda}$ is the solar irradiance at the top of the atmosphere in Wm^{-2}, found in the calibration file of the satellite image, μ is the cosine of the solar zenith angle ($\cos\theta_0$), found on the satellite image header file, τ_r is the Rayleigh optical thickness, and τ_a is the AOT (which is unknown).

The "$\tau(\mu)\uparrow$" [42] is calculated in Eq. (5):

$$t(\mu)\uparrow = \exp\left[-(\tau_r + \tau_a)/\mu\right] \tag{5}$$

Although several equations to calculate τ_r have been developed, most radiative transfer equations used the equation provided by Moller [47], as indicated in Eq. (6), which is considered to be the most appropriate [34]:

$$\tau_r(\lambda_c) = 0.00879\lambda_c^{-4.09} \tag{6}$$

where λ_C is the central wavelength for each Landsat band.

2.1.2. Calculating Rayleigh scattering (L_{pr})

To calculate path radiance due to Rayleigh scattering(L_{pr}), Eq. (7) is used [44]:

$$L_{pr} = \left\{\frac{E_{0\lambda}\, cos\theta_0\, P_r}{4\pi(cos\theta_0 + \cos\theta_v)}\right\}\left\{1 - exp\left[-\tau_r\left({}^1/_{cos\theta_0} + {}^1/_{cos\theta_v}\right)\right]\right\} to_{3\uparrow}\, to_{3\downarrow} \tag{7}$$

where P_r is Rayleigh scattering phase function and θ_v is the sensor viewing angle, as found on the satellite image header file. $to_{3\uparrow}$ refers to the transmittance factor due to ozone in the upward direction, and $to_{3\downarrow}$ refers to the transmittance factor due to ozone in the downward direction, both of which are considered negligible and therefore have a value of 1 [44].

The Rayleigh scattering phase function (P_r) is calculated using Eq. (8) [44]:

$$P_r = \frac{3}{4}\left[+\cos^2(180 - \theta_0)\right] \tag{8}$$

2.1.3. Calculating Mie scattering

To calculate Mie scattering, Eq. (9) is used [47]:

$$L_{pa} = \left\{\frac{E_{0\lambda} cos\theta_0 \omega_a P'_a}{4\pi cos\theta_0 + cos\theta_v}\right\}\left\{1 - exp\left[-\tau_a\left(\left(\frac{1}{cos\theta_0}\right) + \left(\frac{1}{cos\theta_v}\right)\right)\right]\right\} exp\left[-\tau_r\left(\left(\frac{1}{cos\theta_0}\right) + \left(\frac{1}{cos\theta_v}\right)\right)\right] \tag{9}$$

where w_a is the aerosol single scattering albedo as calculated by Waggoner et al. [48] and P'_a is the phase function, as calculated by Gilabert et al. [44], Forster [29], and Turner et al. [45], provided in graph form or downloaded from AERONET stations taken with the Cimel sunphotometer (http://aeronet.gsfc.nasa.gov/).

2.1.4. Steps for performing the image-based integrated method

In order to retrieve AOT using the above equation and solve for τ_a, the below steps should be followed:

Step 1: Download the required Landsat image from the USGS website (glovis.usgs.gov) by selecting the area of interest and the date and time of the satellite overpass.

Step 2: Retrieve the solar zenith angle (θ_0) at the overpass time, according to the header file on the satellite image.

Step 3: Retrieve the sensor viewing angle (θ_v) at the overpass time, according to the header file on the satellite image.

Step 4: Retrieve E_0 according to satellite image calibration file.

Step 5: Calculate μ, which is the cos(θ_0) using the variable derived from step 2.

Step 6: Retrieve λ_C, using satellite image calibration file.

Step 7: Calculate τ_r using Eq. (6), using the variable determined in step 6.

Step 8: Solve for E_G using Eq. (4), using the variables derived from steps 4, 5, and 7. The τ_a (AOT) will be unknown.

Step 9: Find $t(\mu)\uparrow$ using Eq. (5), using the variables derived from steps 5 and 7. The τ_a (AOT) will be unknown.

Step 10: Calculate L_{ts} from the satellite image header file using Eq. (3).

Step 11: Calculate ρ_{tg} from the surface reflectance values.

Step 12: Calculate L_p using Eq. (2), using variables from steps 8–11.

Step 13: Calculate P_r using Eq. (8), using variables from steps 2 and 5.

Step 14: Solve for L_{pr} using Eq. (7), using variables from steps 2, 3, 4, 7, and 13 and using 1 for $to_{3\uparrow}$ and $to_{3\downarrow}$.

Step 15: Enter w_a which is between 0.73 and 0.87 in urban areas and 0.89–0.95 in remote areas.

Step 16: Enter P'_a using phase function graph or from AERONET.

Step 17: Calculate L_{pa} using Eq. (9), using variables from steps 2, 3, 4, 7, 15, and 16.

Step 18: Solve for τ_a (AOT) using Eq. (1), using variables from steps 14 and 17.

2.2. Image-based integrated method for GIS analysis

In order to solve for AOT and conduct a GIS analysis, it may be necessary to simplify the equation due to the complexity of the RT equations and their logarithmic components. The equations for path radiance, Rayleigh scattering, and Mie scattering can be simplified and then combined into one equation, to be solved in an image-based software. After the AOT values are calculated for every pixel in the satellite image, the values can be exported into a GIS geospatial database, with the AOT value and coordinates. In order to display AOT on a GIS thematic map, the Kriging method can be used to interpolate the AOT values and then overlay the data into GIS vector maps or satellite images of the area of interest to display a spatial distribution of AOT. The maps can also be overlaid with vector data including roads, building lots, parks, green areas, stadiums, industrial areas, and municipal boundaries to identify AOT distribution over the area of interest. *In-situ* AOT values, if available within the area of interest, can be correlated with the AOT values from the GIS maps to perform an accuracy assessment for a specific location.

3. AOT retrieval using the image-based integrated method: the example of Limassol, Cyprus

The example described in this chapter includes AOT and surface reflectance field measurements in Limassol Cyprus during Landsat-5 TM and Landsat-7 ETM+ satellite overpass. Only 11 Landsat satellite images with low cloud cover are used. These images were chosen due to their availability and clarity. The Landsat-5 TM images used are for 24/6/2010, 10/7/2010, 27/8/2010, and 28/9/2010, and the Landsat-7 ETM+ images used are for 13/4/2010, 29/4/2010, 31/5/2010, 16/6/2010, 7/11/2010, 9/12/2010, and 2/5/2011. During each satellite overpass, AOT measurements were taken using the Microtops II sunphotometer at five different locations in Limassol and the stationery Cimel sunphotometer, which is part of the AERONET network, located in the center of the study area. These AOT values are later compared to the AOT values derived from the algorithm using the satellite images. At the same time, the SVC HR-1024 spectroradiometer was used to measure the reflectance value of a large nonvariant target dark asphalt surface located near the center of the satellite image.

Geometric and radiometric correction is performed on the 11 Landsat-5 and Landsat-7 satellite images, as well as atmospheric correction using the modified DP method. The image-based integrated method is applied to the Landsat-5 and Landsat-7 images taken over specific days of 2010–2011. Eqs. (1–9) are used to retrieve AOT from bands 1–4 of the Landsat-5 and Landsat-7 satellite images, following the steps outlined. As the *in-situ* AOT measurements are at the 500 nm wavelength, Landsat band 1 is used to derive AOT levels in this method.

In this example, the model is run in the ERDAS Imagine and MATLAB software to generate an image that is consists of AOT values. The model is then applied to every surface using the Landsat band 1 satellite image, since the AOT required are in the 500 nm wavelength. The resulting image features the AOT values for every pixel in the image, as featured in **Figure 2**. Areas with no data, due to cloud presence or negative AOT values, appear in white. All AOT values can be exported to ASCII files to create a georeferenced AOT dataset, which is then imported into the ArcGIS software.

Using the ArcGIS software, each pixel is converted to a point and each point is associated with the calculated AOT value. The white area of the image contains values of "no data" (**Figure 3**, left). To create a GIS thematic map showing the AOT distribution in the urban area of Limassol, the interpolation method is used in order to estimate the values from the "no data" regions. The ordinary Kriging interpolation tool is applied, where the unknown values of the "no data" area are interpolated using the weighted average of neighboring samples. **Figure 3** (right) shows the results of the Kriging interpolation for the Limassol example, where the dots indicate the points with AOT values, while the colored sections are the resulting AOT values determined from Kriging interpolation.

Figure 2. AOT values derived from algorithm.

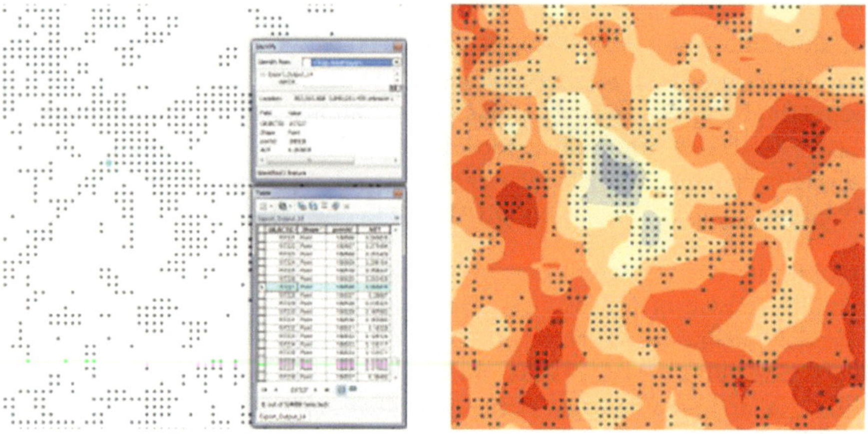

Figure 3. (Left) Points with AOT dataset. (Right) Interpolation with Kriging method.

Once the interpolation is completed, an AOT thematic map is created, which is classified according to AOT values. The AOT values are displayed in different colors according to the AOT concentration and a legend of the AOT values is created. This facilitates the ability to visualize the concentration and distribution of AOT values over an urban area. **Figure 4** features a map of Limassol following Kriging interpolation, which provides synoptic views of the AOT distribution. As is evident, high AOT levels are present throughout the city, especially in the industrial estates and on busy streets, with lower AOT levels present in parks, stadiums, and outside the city.

http://dx.doi.org/10.5772/65279

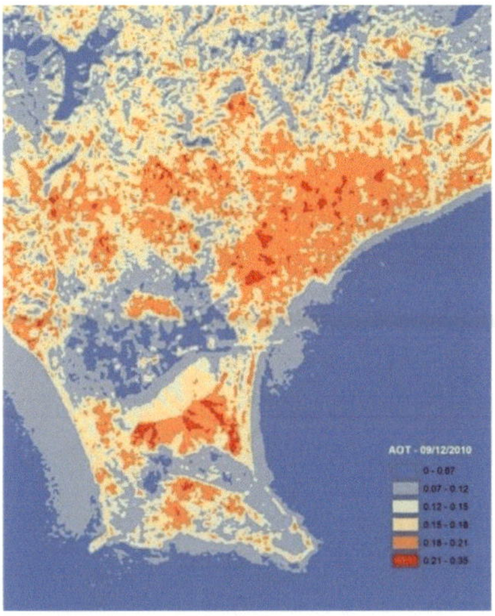

Figure 4. Thematic map of Limassol (09/12/2010) after Kriging interpolation, with AOT values.

In the example, the thematic map is overlaid with GIS vector data from the Lands and Surveys Department for Limassol to facilitate the identification of sources of AOT values within the area of interest (**Figure 5**). Each polygon in the thematic map is connected to a GIS database, where data such as plot number and area information can be identified. In this way, the GIS database, combined with the AOT data, can create thematic maps to illustrate trends in different areas.

Figure 5. Vector overlay.

In the example, the image-based integrated method is used to produce GIS maps indicating the AOT distribution over Limassol. In order to visualize high and low AOT levels, thematic maps are generated using colors ranging from blue to green to yellow to red for the specified

AOT range. A separate thematic map is created for each of the 11 Landsat-5 and Landsat-7 satellite images. **Figure 6** features the GIS thematic map using Landsat-7 satellite imagery from 24/6/2010.

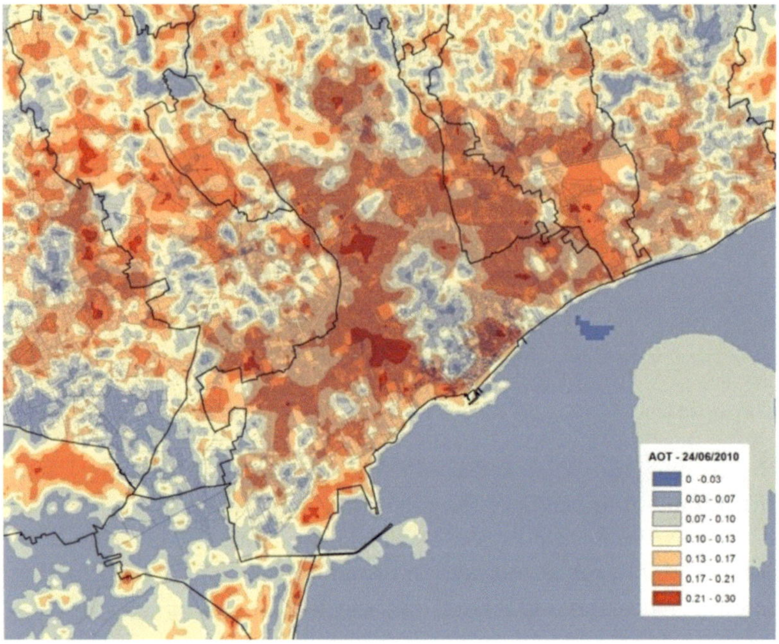

Figure 6. GIS thematic map, 24 June 2010, with vector overlay.

The AOT values derived from the algorithm are compared with the AOT values measured with the Microtops and Cimel sun photometers. **Figure 7** shows all the locations of the *in-situ* AOT measurements using sun photometers. Locations 1–5 are measured with the Microtops sunphotometer, while location 6 is measured with the Cimel sunphotometer. In **Figure 7**, location 7 refers to the Department of Labor Air Quality Monitoring Site, which was used only for reference. An accuracy assessment is done to compare the AOT from the GIS map with the *in-situ* AOT as measured from the Microtops and Cimel sunphotometers during satellite overpass. A linear regression is conducted of the AOT levels measured with the sunphotometers and the AOT values derived from the algorithm, with a coefficient of determination $R2$ of 0.977.

Table 1 compares the AOT measurements taken on site using a sunphotometer against the AOT values derived from the GIS maps using the proposed algorithm by location and date, as well as the correlation coefficient (r) between the on-site AOT measurements and algorithm-derived AOT values from GIS, for each location. In order to determine the accuracy of the GIS model, the root mean square deviation (RMSD) is also calculated between the on-site AOT measurements and algorithm-derived AOT values from GIS for each location. The results of the RMSD, correlation coefficients, and the coefficient of determination verify that the AOT derived from the GIS map correlate strongly with the on-site AOT values as measured by sunphotometers.

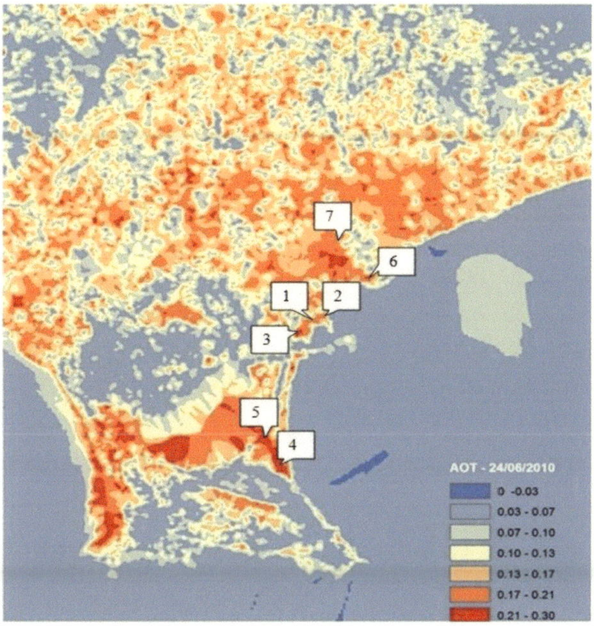

Figure 7. Location of *in-situ* AOT measurements.

	Site 1		Site 2		Site 3	
	AOT measured	AOT algorithm	AOT measured	AOT algorithm	AOT measured	AOT algorithm
13-Apr	0.265	0.261	0.263	0.261	0.257	0.256
29-Apr	0.328	0.320	0.435	0.400	0.310	0.315
31-May	0.269	0.268	0.271	0.272	0.449	0.400
16-Jun	0.290	0.292	0.295	0.296	0.324	0.325
24-Jun	0.292	0.293	0.294	0.295	0.288	0.290
10-Jul	0.271	0.270	0.339	0.350	0.339	0.350
27-Aug	0.282	0.280	0.333	0.340	0.320	0.325
28-Sep	0.208	0.200	0.235	0.234	0.200	0.202
7-Nov	0.235	0.238	0.221	0.225	0.231	0.235
9-Dec	0.200	0.205	0.232	0.228	0.228	0.230
2-May	0.313	0.310	0.355	0.355	0.349	0.350
r	.994		.985		.98	
RMSD	.004		.002		.001	

	Site 4		Site 5		Site 6	
	AOT measured	AOT algorithm	AOT measured	AOT algorithm	AOT measured	AOT algorithm
13-Apr	0.254	0.252	0.260	0.258	0.256	0.252
29-Apr	0.393	0.390	0.391	0.390	0.421	0.400
31-May	0.360	0.359	0.321	.0320	0.186	0.220
16-Jun	0.344	0.345	0.321	0.322	0.291	0.290
24-Jun	0.299	0.300	0.290	0.291	0.118	0.120
10-Jul	0.270	0.270	0.341	0.350	0.354	0.350
27-Aug	0.347	0.350	0.333	0.335	0.322	0.325
28-Sep	0.235	0.240	0.263	0.265	0.195	0.200
7-Nov	0.290	0.295	0.232	0.240	0.173	0.200
9-Dec	0.244	0.245	0.260	0.263	....	0.210
2-May	0.248	0.250	0.283	0.285	....	0.330
r	0.999		.997		.931	
RMSD	0.0025		0.002		.004	

Table 1. Comparison of on-site AOT and AOT from GIS algorithm (4/2010–5/2011).

4. Conclusions

In this chapter, the image-based integrated method is developed to retrieve AOT values from satellite images and display these values in an urban area, using the example of Limassol, Cyprus. The findings show that the proposed integrated image-based method is able to accurately obtain AOT measurements from satellite images. The accuracy assessment shows strong agreement between the *in-situ* AOT values and the AOT values retrieved through the method described in the chapter. The proposed methodology is able to retrieve AOT using satellite images and visually display AOT over urban areas by using GIS to produce thematic maps. Also, the significance of the image-based integrated method is that the environmental data is not needed, as compared with other methods. This methodology is therefore an alternative to more sophisticated and complex methods of deriving AOT values from satellite images, especially where atmospheric or meteorological data are not accessible. Also, the methodology can be used with archived satellite images where environmental data are not available, thereby providing detailed information regarding spatial aerosol concentration overtime. Also, thematic maps can be generated to illustrate the distribution of AOT overtime in a specific area of interest and enabling trends to be identified.

Acknowledgements

The author would like to thank the Cyprus University of Technology, Department of Civil Engineering and Geomatics, Remote Sensing Lab for use of the sunphotometers and AERONET data.

Author details

Kyriacos Themistocleous

Address all correspondence to: k.themistocleous@cut.ac.cy

Department of Civil Engineering and Geomatics, Cyprus University of Technology, Limassol, Cyprus

References

[1] Anderson, T., Charlson, R., Schwartz, S., Knutti, R., Boucher, O., Rodhe, H. & Heintzenberg, J. Climate forcing by aerosol—a hazy picture. *Science*. 2003;300:1103–1104.

[2] Kaufman, Y.J., Tanre, D. & Boucher, O. A satellite view of aerosols in the climate system. *Nature*. 2002;419:215–223. DOI: 10.1038/nature 01091

[3] Holben, B.N., Smirnov, A., Eck, T.F., Slutsker, I., Abuhassan, N., Newcomb, W., Schafer, J.S., Tanre, D., Chatenet, B. & Lavenu, F. An emerging ground-based aerosol climatology: aerosol optical depth from AERONET. *Journal of Geophysical Research*. 2001;106:12067–12098.

[4] Yu, H., Dickinson, R.E., Chin, M., Kaufman, Y.J., Zhou, L., Tian, Y., Dubovik, O. & Holben, B.N. Direct radiative effect of aerosols as determined from a combination of MODIS retrievals and GOCART simulations. *Journal of Geophysical Research*. 2004;109:1–13(D03206). DOI: 10.1029/2003JD003914

[5] Remer, L., Kaufman, Y.J. & Tanre, D. The MODIS aerosol algorithm, products, and validation. *Journal of the Atmospheric Sciences*. 2005;62(4):947–973.

[6] Misra, A., Jayaraman, A. & Ganguly, D. Validation of MODIS derived aerosol optical depth over western India. *Journal of Geophysical Research*. 2008;113:1–12 DOI: 10.1029/2007JD009075

[7] Guleria, R.P., Kuniyal, J.C., Rawat, P.S., Thakur, H.K., Sharma, M., Sharma, N.L., Singh, M., Chand, K., Sharma, P., Thakur, A.K., Dhyani, P.P. & Bhuyan, P.K. Aerosols optical properties in dynamic atmosphere in the northwestern part of the Indian Himalaya: a comparative study from ground and satellite based observations. *AtmosphericResearch*. 2011;101(3):726–738.

[8] Retalis, A., Cartalis, C. & Athanassiou, E. Assessment of the distribution of aerosols in the area of Athens with the use of LANDSAT Thematic Mapper data. *International Journal of Remote Sensing*. 1999;20(5):939–945.

[9] Hadjimitsis, D.G. Aerosol optical thickness (AOT) retrieval over land using satellite image-based algorithm. *Air Quality Atmosphere and Health-An International Journal*. 2009;2(2):89–97.

[10] Huang, D., Knyazikhin, Y., Wang, W., Deering, D.W. & Stenberg, P. Stochastic transport theory for investigating the three-dimensional canopy structure from space measurements. *Remote Sensing of Environment*. 2008;112:35–50.

[11] Themistocleous, K. & Hadjimitsis, D.G. Development of an image-based integrated method for determining andmapping aerosol optical thickness (AOT) over urban areas using the darkestpixel atmospheric correction method, RT equation and GIS: a case study ofthe Limassol area in Cyprus. *ISPRS Journal of Photogrammetry and Remote Sensing*. 2013;86:1–10. DOI: 10.1016/j.isprsjprs.2013.08.011

[12] Zawadzka, O., Markowicz, K.M., Pietruczuk, A., Zielinski, T. & Jaroslawski, J. Impact of urban pollution emitted in Warsaw on aerosol properties. *Atmospheric Environment*. 2013;69:15–28.

[13] Yahi, H., Santer, R., Weill, A., Crepon, M. & Thiria, S. Exploratory study for estimating atmospheric low level particle pollution based on vertical integrated optical measurements. *Atmospheric Environment*. 2011;45(23):3891–3902.

[14] Wang, Z., Chen, L., Tao, J., Zhang, Y. & Su, L. Satellite-based estimation of regional particulate matter (PM) in Beijing using vertical-and-RH correcting method. *Remote Sensing of Environment*. 2010;114(15):50–63.

[15] Jiang, X., Liu, Y., Yu, B. & Jiang, M. Comparison of MISR aerosol optical thickness with AERONET measurements in Beijing metropolitan area. *Remote Sensing of Environment*. 2007;107:45–53.

[16] Hadjimitsis, D.G. & Clayton, C.R.I. Detecting air pollution from space using image-based method. In: *International Conference: Protection and Restoration of the Environment*, VIII, 313; Crete, Greece. 2006.

[17] Koukouli, M.E., Balis, D.S., Amiridis, V., Kazadzis, S., Bais, A., Nickovic, S. & Torres, O. Aerosol variability over Thessaloniki using ground based remote sensing observations and the TOMS aerosol index. *Atmospheric Environment*. 2006;40(28):5367–5378.

[18] Hadjimitsis, D.G., Retalis, A. & Clayton, C.R.I. The assessment of atmospheric pollution using satellite remote sensing techniques in large cities in the vicinity of airports. *Water, Air and Soil Pollution: Focus, An International Journal of Environmental Pollution*. 2002;2(5–6):631–640.

[19] Tang, J., Xue, T., Yuc, T. & Guen, Y.N. Aerosol optical thickness determination by exploiting the synergy of TERRRA and AQUA MODIS. *Remote Sensing and Environment*. 2005;94:327–334.

[20] Kanaroglou, P.S., Soulakellis, N.A. & Sifakis, N.I. Improvement of satellite derived pollution maps with the use of a geostatistical interpolation method. *Journal of Geographical Systems*. 2002;4:193–208.

[21] Prospero, J.M., Ginoux, P., Torres, O., Nicholson, S. & Gill, T. Environmental characterization of global sources of atmospheric soil dust identified with the NIMBUS7 total ozone mapping spectrometer (TOMS) absorbing aerosol product. *Review of Geophysics*. 2002;40(1):1002. DOI: 10.1029/2000RG000095

[22] Sifakis, N. & Deschamps, P.Y. Mapping of air pollution using SPOT satellite data. *Photogrammetric Engineering and Remote Sensing*. 1992;58:1433–1437.

[23] Leon, J., Chazette, P. & Dulac, F. Direct and indirect methods for correcting the aerosol effect on remote sensing. *Remote Sensing and Environment*. 1995;55:65–79.

[24] Sifakis, N., Soulakellis, N. & Paronis, D. Quantitative mapping of air pollution density using Earth observations: a new processing method and application to an urban area. *International Journal of Remote Sensing*. 1998;19(17):3289–3300.

[25] Guanter, L., Gómez-Chova, L. & Moreno, J. Coupled retrieval of aerosol optical thickness, columnar water vapor and surface reflectance maps from ENVISAT/MERIS data over land. *Remote Sensing of Environment*. 2008;112:2898–2913.

[26] Hadjimitsis, D.G., Themistocleous, K., Vryonides, P., Toulios, L. & Clayton, C.R.I. Applications of satellite remote sensing and GIS to urban air-quality monitoring: potential solutions and suggestions for the Cyprus area. In: *6th International Conference on Urban Air Quality*, 144; Limassol, Cyprus. 2007.

[27] Hadjimitsis, D.G. & Clayton, C.R.I. Determination of the aerosol optical thickness and assessment of atmospheric conditions using satellite image-based processing algorithm and radiative transfer theory. In: *7th Pan-Hellenic International Conference of Meteorology, Climatology and Atmospheric Physics*; University of Cyprus, Nicosia, Cyprus. 2004.

[28] Hadjimitsis, D.G., Clayton, C.R.I. & Hope, V.S. An assessment of the effectiveness of atmospheric correction algorithms through the remote sensing of some reservoirs. *International Journal of Remote Sensing*. 2004;25(18):3651–3674.

[29] Forster, B.C. Derivation of atmospheric correction procedures for Landsat MSS with particular reference to urban data. *International Journal of Remote Sensing*. 1984;5(5): 799–817.

[30] Istomina, L.G., von Hoyningen-Huene, W., Kokhanovsky, A.A., Schultz, E. & Burrows, J.P. Remote sensing of aerosols over snow using infrared AATSR observation. *Atmospheric Measurement Techniques*. 2011;4:1133–1145. DOI: 10.51941.amt-4-1133-2011

[31] Lee, K.H. & Kim, Y.J. Satellite remote sensing of Asian aerosols: a case study of clean, polluted and dust storm days. *Atmospheric Measurement Techniques Discussions*. 2010;3:2651–2680. DOI: 10.5194/amtd-3-2651-2010

[32] Xue, Y., Wan W., Li, Y.J., Guang, J., Bai, L.Y., Wang Y. & Ai, J.W. A data intensive scientific computing framework for quantitative retrieval of geophysical parameters using satellite data. *IEEE Computer*. 2008;41(4):33–40.

[33] Seidel, F., Schlapfer, D., Nieke, J. & Itten, K.I. Sensor performance requirements for the retrieval of atmospheric aerosols by airborne optical remote sensing. *Sensors*. 2008;8:1901–1914.

[34] Hadjimitsis, D.G. The application of atmospheric correction algorithms in the satellite remote sensing of reservoirs [dissertation]. Guildford, U.K.: University of Surrey, School of Engineering in the Environment, Department of Civil Engineering; 1999.

[35] Kaufman, Y.J. & Sendra, C. Algorithm for automatic atmospheric corrections to visible and near-IR satellite imagery. *International Journal of Remote Sensing*. 1988;9(8):1357–1381.

[36] Kaufman, Y.J. & Tanre, D. Direct and indirect methods for correcting the aerosol effect on remote sensing. *Remote Sensing of Environment*. 1996;55:65–79.

[37] Kaufman, Y.J., Tanre, D., Remer, L.A., Vermote, E.F., Chu, A. & Holben, B. N. Operational remote sensing of tropospheric aerosol over land from EOS moderate resolution imaging spectroradiometer. *Journal of Geophysical Research*. 1997;102:17051–17067.

[38] Schott, J.R. *Remote Sensing: The Image Chain Approach*. Oxford, UK: Oxford University Press; 2007.

[39] Cheng, K.S. & Lei, T.C. Reservoir trophic state evaluation using Landsat TM images. *Journal of the American Water Resources Association*. 2001;37:1321–1334.

[40] Hadjimitsis, D.G. & Clayton, C.R.I. Assessment of temporal variations of water quality in inland water bodies using atmospheric corrected remotely sensed image data. *Environmental Monitoring and Assessment*. 2009;159(1–4):281–292.

[41] deHaan, J.F., Hovenier, J.W., Kokke, J.M.M. & Van Stokkom, H.T.C. Removal of atmospheric influences on satellite-borne imagery: a radiative transfer approach. *Remote Sensing of Environment*. 1991;37:1–21.

[42] Hill, J. & Sturm, B. Radiometric correction of multi-temporal Thematic Mapper data for use in agricultural land cover classification and vegetation monitoring. *International Journal of Remote Sensing*. 1991;12(7):1471–1491.

[43] Deschamps, P.Y., Herman, M. & Tanré, D. Modeling of the atmospheric effects and its application to the remote sensing of ocean colour. *Applied Optics*. 1983a;23:3751–3758.

[44] Gilabert, M.A., Conese, C. & Maselli. An atmospheric correction method for the automatic retrieval of surface reflectances from TM image. *International Journal of Remote Sensing*. 1994;15(10):2065–2086.

[45] Turner, R.E. & Spencer, M.M. Atmospheric model for correction of spacecraft data. In: *Eight International Symposium on Remote Sensing of the Environment*; 895–934; Ann Arbor, Michigan; 1972.

[46] Landsat-7 Science Data Users Handbook. National Aeronautics and Space Administration; 2011. Accessed at: http://landsathandbook.gsfc.nasa.gov/pdfs/Landsat7_Handbook.pdf

[47] Moller, F. *Strahlung in der UnterenAtmosphare, Handbuch der Physik*. New York: Springer-Verlag; 1957.

[48] Waggoner, A.P., Weiss, R.E., Ahlquist, N.C., Covert, D.S., Wills, S. & Charlson, R.J. Optical characteristics of atmospheric aerosols. *Atmospheric Environmen*t. 1981;15:1891–1909.

Chapter 7

Spatial Distribution of Aerosol Optical Thickness Retrieved from SeaWiFS Images by a Neural Network Inversion over the West African Coast

Daouda Diouf, Awa Niang, Julien Brajard,
Salam Sawadogo, Michel Crepon and Sylvie Thiria

Additional information is available at the end of the chapter

http://dx.doi.org/10.5772/65874

Abstract

Aerosol optical thickness (AOT) was provided by SeaWiFS over oceans from October 1997 to December 2010. Weekly, monthly, and annually maps might help scientifics to better understand climate change and its impacts. Making average of several images to get these maps is not suitable on West African coast. A particularity of this area is that it is constantly traversed by desert dust. The algorithm used by SeaWiFS inverts the reflectance measurements to retrieve the aerosol optical thickness at 865 nm. For the poorly absorbing aerosol optical thickness less than 0.35, the standard algorithm works very well. On the west African coast that is often crossed by desert aerosol plumes characterized by high optical thicknesses. In this paper we study the spatial and temporal variability of aerosols on the West African coast during the period from December 1997 to November 2009 by using neural network inversion. The neural network method we used is mixed method of neuro-variational inversion called SOM-NV. It is an evolution of NeuroVaria that is a combination of a variational inversion and multilayer perceptrons, multilayer perceptrons (MLPs). This work also enables validation of the optical thickness retrieved by SOM-NV with AOT in situ measurements collected at AErosol RObotic NETwork (AERONET) stations.

Keywords: West African coast, absorbing aerosols, neural networks

1. Introduction

The sensor SeaWiFS ocean color provides daily luminance measurements of ocean-atmosphere system in the visible and near infrared since October 1997. Luminances are at wavelengths 412,

443, 490, 510, 555, 670, 765, and 865 nm. For each wavelength λ, the TOA reflectance ρ is computed as

$$\rho(\lambda, \theta_V, \phi) = \frac{\pi \cdot L(\lambda, \theta_V, \Delta\Phi)}{E_0(\lambda) \cdot \cos(\theta_S)} \tag{1}$$

where $E_0(\lambda)$ is the extraterrestrial solar irradiance (in Wm^{-2} nm^{-1}, varying with the sun-earth distance), θ_s and θ_v are the sun- and satellite-viewing zenith angles, respectively, and $\Delta\Phi = \phi_o - \phi_v$ is the azimuth angle difference between the satellite and the sun.

SeaWiFS aerosol products are generated, validated, and made available by NASA. These aerosols from the standard atmospheric correction algorithm can hardly be used for global aerosol studies because of aerosol optical thickness greater than about 0.35.

Plumes of absorbing aerosols are observed regularly on the West African coast near the Sahara, which prevents spatial and temporal operation of atmospheric and oceanic parameters.

This paper presents a new method to retrieve the aerosol parameters from ocean color satellite radiometer and is able to give information on absorbing aerosols.

2. Data and method

For this study, we use daily luminance measurements made by the SeaWiFS sensor off the West African coast in an area between 8°–24°N and 14°–30°W. These measures extend the period of 1997–2009. The TOA reflectance ρ is the sum of contributions to the signal of each constituent of the atmosphere and ocean. The contribution of Rayleigh scattering, specular reflection, and absorption gas is known a priori with accuracy and has been removed from the signal. The pixels of clouds have been removed using the cloud detection algorithm [1]. Thus, the corrected reflectance is

$$\rho_{used} = \rho_a + \rho_{ra} + t\rho_w \tag{2}$$

where $\rho_a + \rho_{ra}$ is the atmospheric contribution, ρ_w is the contribution of the water, and t is the transmittance of the atmosphere at a given wavelength (λ).

We used satellite data sets comprising ten-dimensional vectors, whose components are eight wavelengths measured by the radiometer and two viewing angles since the reflectance spectra depend on the geometry of the measurement. These angles are the sun zenith angle θ_s, and the scattering angle γ is defined as

$$\gamma = \arccos(-\cos\theta_v \cos\theta_s + \sin\theta_v \sin\theta_s \cos\Delta\Phi) \tag{3}$$

Each vector, whose components correspond to the SeaWiFS wavelengths, represents a ρ_{used} spectrum.

http://dx.doi.org/10.5772/65874

The learning dataset $Data^{obs}$ consists of ρ_{used}^{obs} observed at eight wavelengths (412, 443, 490, 510, 555, 670, 765, and 865 nm) extracted from pixels of SeaWiFS images during the year 2003 and two associated viewing angles (i.e., the sun zenith angle θ_s and the scattering angle γ). $Data^{obs}$ is thus composed of ten-component vectors.

We also used theoretical database $\mathrm{Data}^{\mathrm{expert}}$ that consists of the ρ_{used}^{expert} computed at eight wavelengths with a 2-layer radiative transfer model [2] for various optical thickness values, chlorophyll content, and geometry of the measurement and for five aerosol models.

Each $\mathrm{Data}^{\mathrm{expert}}$ vector comprises eight spectral components (ρ_{used}^{expert}) and two geometry components which are the sun zenith angle θ_s and the scattering angle γ. $\mathrm{Data}^{\mathrm{expert}}$ comprises 12,000,000 simulated vectors using four aerosol models (maritime, oceanic, coastal, and tropospheric) [3] and one absorbing aerosol (African dusts) [1]. The five aerosol models were computed at four different relative humidity (70%, 80%, 90%, and 99%).

$\mathrm{Data}^{\mathrm{expert}}$ was used in order to introduce the expertise and to retrieve the aerosol type and the optical thickness values.

In this study, we used SOM-NV [2], a two successive statistical models for analyzing the $Data^{obs}$ images: the self-organizing map (SOM) [4] model and the NeuroVaria method [5, 6]. We first processed the images with a SOM model, which is well suited for visualizing and clustering a high-dimensional data set. We denoted this topological map as SOM-Angle-Spectrum (SOM-A-S). In the light of the results obtained by Niang et al. [7], we chose a similar architecture for SOM-A-S, a two-dimensional array with a large number of neurons (20 × 30 = 600). SOM-A-S was learned on the $Data^{obs}$ of the year 2003. The vectors of the learning data set were thus clustered into 600 groups, allowing a highly discriminative representation of $Data^{obs}$. The second dataset, $Data^{expert}$, representing the expertise, was used to decode the SeaWiFS images. The principle of the method is to compare the ten-component vectors of $Data^{expert}$ whose associated parameters are known, with those of the neurons of SOM-A-S according to a distance. At the end of the labeling, each neuron of SOM-A-S map has captured a set of ρ^{expert} and takes a label, which is extracted from that set according to the procedure described in [7]. Each neuron is therefore associated with an atmospheric and ocean physical parameters (τ (AOT: aerosol optical thickness), C (*chlorophyll-a pigment*)), and an aerosol type. The SOM-A-S map being labeled, we are able to analyze a satellite image by projecting the ten-component vector (reflectances and viewing angles) associated with each pixel on the SOM-A-S map. Pixels captured by a neuron are assigned to the aerosol type and optical thickness associated with this neuron. For monthly climatology images, the aerosol type is estimated as the median of the types of the images considered.

The second statistical model improves the retrieval of the optical thickness. We used a neuro-variational algorithm called NeuroVaria that is able to provide accurate atmospheric corrections for inverting satellite ocean color measurements. The algorithm minimizes a weighted quadratic cost function, J, by adjusting control parameters (atmospheric and oceanic) such as τ and C [8]. J describes the difference between the satellite measurement $\rho_{toa}^{mes}(\lambda)$ ρ^{obs} and a simulated reflectance $\rho_{toa}^{sim}(\lambda)\rho^{sim}$ computed using radiative transfer codes modeled by supervised neural networks (the so-called multilayer perceptrons, MLPs). The minimization implies the computation of the gradient of *J* with respect to the control parameters and consequently of the derivatives of the MLPs, which is done by the classical gradient back-propagation algorithm [9]. On this version of SOM-NV, the MLPs modeling the radiative transfer codes were specially designed to take African dusts into account.

A major advantage of the method is a gain in number of processed pixels from SeaWiFS. This work also enables validation of the optical thickness retrieved by SOM-NV with in situ measurements of optical thicknesses AERONET collected at stations in Dakar and Cabo Verde [2].

The complete methodology was applied to SeaWiFS images of the ocean off the West African coast from 1997 to 2009 to produce the type of aerosol, the aerosol optical thickness, and the chlorophyll-a concentration.

Monthly mean map aerosols and chlorophyll-a were calculated on 9 km × 9 km used for SeaWiFS GAC product level 3. Seasonality strong $\tau(865)$ is characterized by a strong invasion of dust into the months of June, July, and August. Their intensities vary from year to year, depending on aridity conditions in Africa and the wind direction.

3. Results

In **Figure 1**, we compare for each pixel of the image (March 10, 2004), the optical thickness and *Chl-a* concentration given by SeaWiFS algorithm (column 2) and those retrieved with SOM-NV (column 1). More than half are fully covered by Saharan dust with high optical thickness in these parts. In most standard SeaWiFS algorithm, a wide area of the image is not processed, while the satellite image can clearly observe the absorbing aerosol event. The SeaWiFS algorithm does not

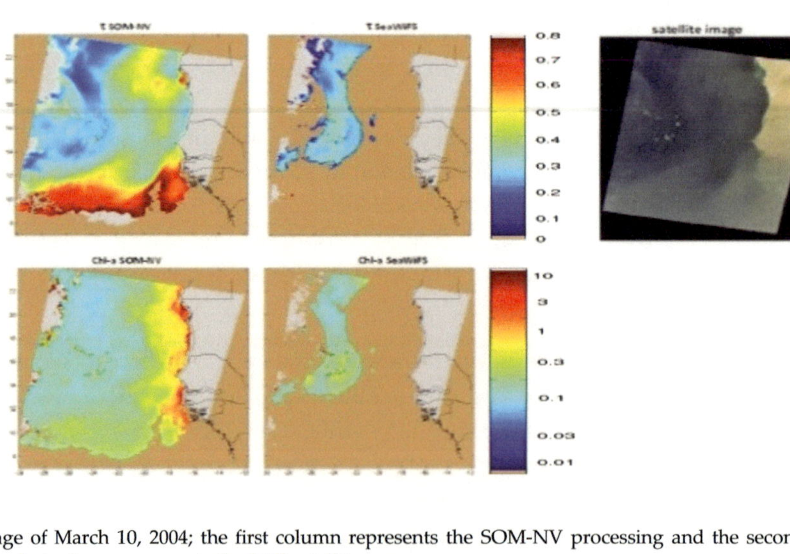

Figure 1. Image of March 10, 2004; the first column represents the SOM-NV processing and the second the SeaWiFS processing. The last column represents the RGB satellite image.

generally process Saharan dust (see image in "true" colors); these absorbing Saharan dust are generally characterized by high optical thickness. The SOM-NV method permits retrieval of the aerosol type and aerosol optical properties from the statistical properties of the data. These retrievals are accurate for optical thickness values higher than 0.35. This is not the case for the standard SeaWiFS product. This permitted us to dramatically increase the number of pixels processed with respect to the standard SeaWiFS algorithm by an order of magnitude.

Maps of average aerosol and chlorophyll-a were computed on grids of 9 km × 9 km GAC used for SeaWiFS level-3 GAC products. The averages obtained by SOM-NV seem statistically more representative than those obtained by SeaWiFS. Monthly averages of optical thickness obtained by the standard processing for the months of March and June of **Figure 2** are drastic because only τ values not exceeding 0.35 are taken into account. These failures demonstrate impossibility for SeaWiFS to establish spatial and temporal global maps of aerosol particularly in absorbing area. This is due to the fact that desert aerosols frequently cross the ocean [10] preventing the standard algorithm to retrieve these aerosols and chlorophyll-a below them.

In **Figure 3**, average of AOT (top), Chl-a (middle) of the months of January and April 2005 is shown. As seen in **Figure 2**, we note also in **Figure 3** failure for SeaWiFS to establish consistent spatial and temporal global maps of aerosol optical thickness. Fluctuations in the chlorophyll-a observed at **Figure 3** are the same in the image for both algorithms. In terms of intensity, the values of Chl-a are higher for the standard algorithm at the continental shelves. An intercomparison with the SeaWiFS standard processing showed that the SOM-NV methodology increased the number of pixels processed of a factor until 10. This is due to the impossibility of the standard algorithm to retrieve the chlorophyll-a in the presence of Saharan dust, which frequently cross the ocean. The low number of pixels processed by SeaWiFS may bias the mean chlorophyll-a maps.

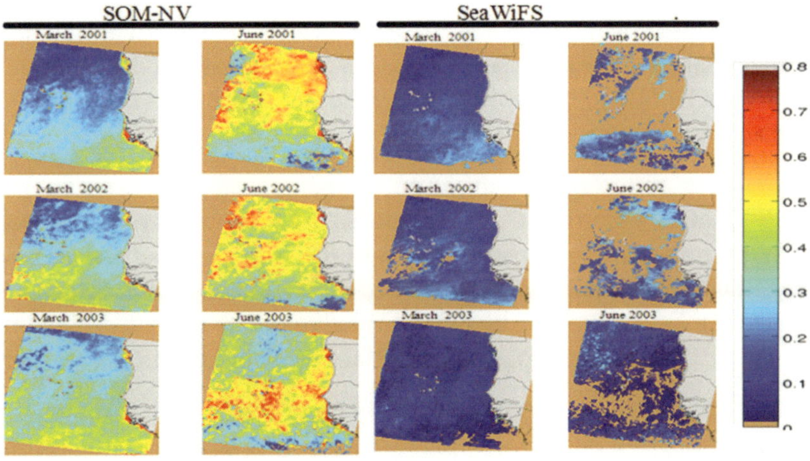

Figure 2. Spatial distribution of the aerosol optical thickness. Monthly average in March and June of the years 2001, 2002, and 2003. We note the same trends from 1 year to another. The aerosol optical thickness are strong in June and relatively low during the month of March for the SOM-NV processing (*left block*) and very low for the standard processing (*right block*).

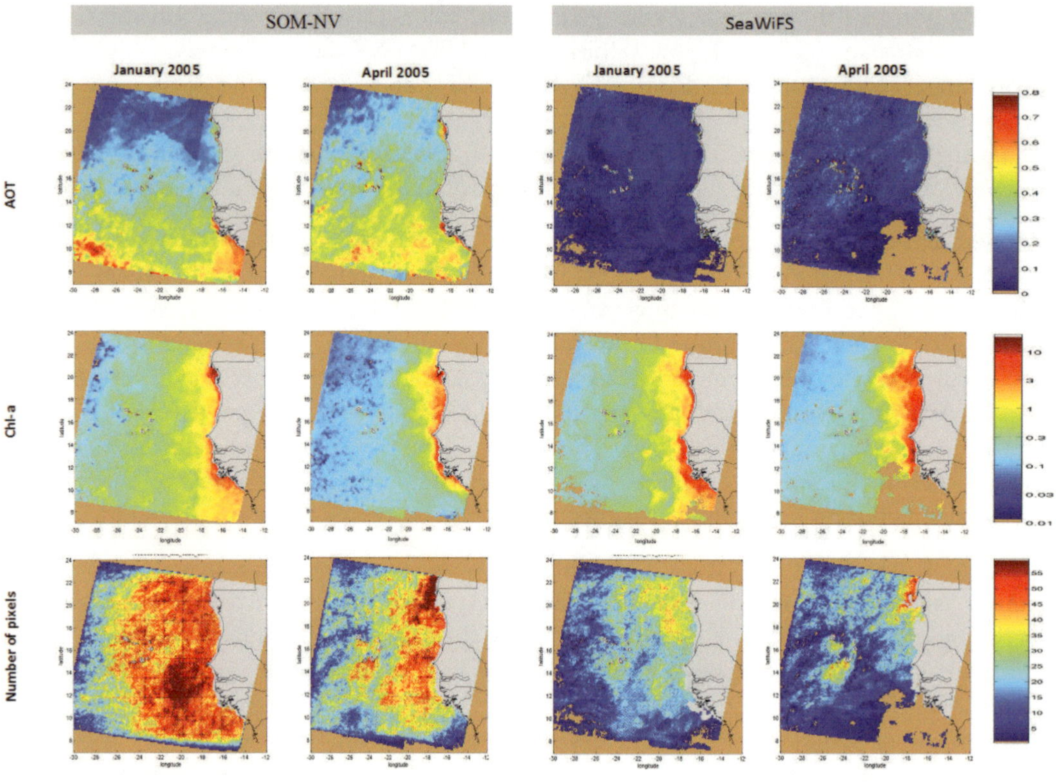

Figure 3. Average of AOT (*top*), Chl-a (*middle*) of the months of January and April 2005 and the number of pixels used in calculating the average (*bottom*).

The presence of the case 2 waters explains these high values of Chl-a. The hypothesis of dark ocean is no longer valid in case of water 2. These results confirm that atmospheric corrections on the Saharan dust by SOM-NV are correct because Chl-a retrievals using OC4V4 are satisfactory spatial and in intensity.

Direct measurements of the optical thickness at 865 nm $\tau(865)$ were performed at two ground stations (Dakar-M'Bour—14°24N and 16°58W and Cabo Verde-Sal Island—16°45N and 22°57W). These measurements were made in the framework of the AErosol RObotic NETwork (AERONET) program which is a federated international network of sun/sky radiometers [11, 12]. Level 2.0 sun photometer measurements $Data^{photometer}$ (cloud screened and quality assured) at the two ground stations are available from 1998 to 2009.

The two AERONET measurements are found when at least one half of the SeaWiFS, or the SOM-NV retrievals within a 5 × 5 pixels square box containing the AERONET site are valid. The AERONET optical thickness values used for this validation were actually the mean of all the measurements made between 12:00 UT and 14:00 UT, because SeaWiFS images over the area were taken around 13:00 UT. We computed mean value for each month.

http://dx.doi.org/10.5772/65874

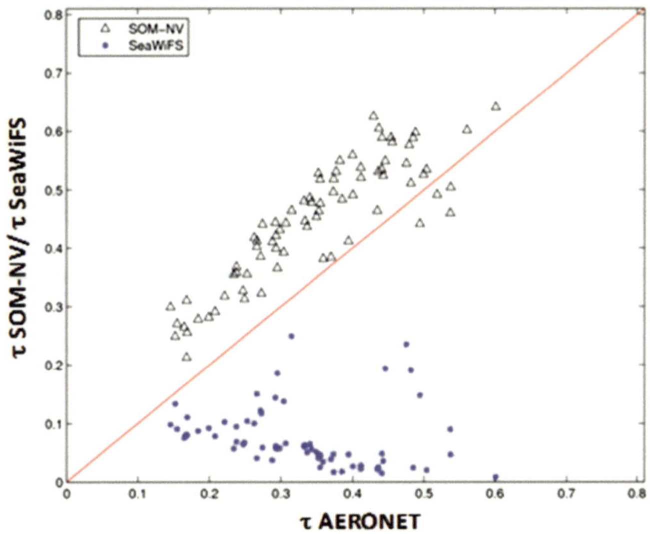

Figure 4. Scatter plot between monthly mean optical thickness measurements computed by SOM-NV (Δ) and the SeaWiFS product (*) and the AERONET measurement at Dakar.

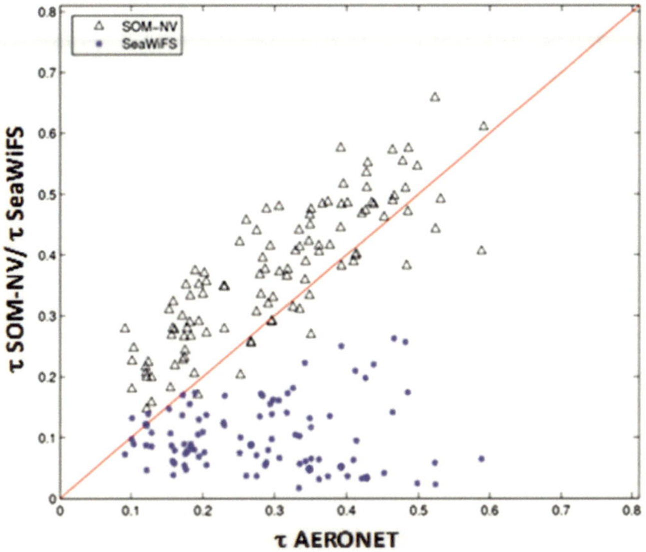

Figure 5. Scatter plot between monthly mean optical thickness measurements computed by SOM-NV (Δ) and the SeaWiFS product (*) and the AERONET measurement at Cabo Verde.

	SOM-NV	SeaWiFS
Mean relative error	38.8%	74.54%
Correlation	80.23	13.45%

Table 1. The performance indicators in Dakar site.

	SOM-NV	SeaWiFS
Mean relative error	40.5%	74.75%
Correlation	82.34%	6.38%

Table 2. The performance indicators in Cabo Verde site.

Figures 4 and **5** confirm that the mean aerosol maps shown above in **Figure 2** are better for SOM-NV than SeaWiFS.

We compared the root-mean-square error (RMSE) and the mean relative error (MRE) of $\tau^{SOM\text{-}NV}$ and τ^{SeaW} with respect to the observed τ^{AERO}.

The MRE remains low for SOM-NV (38.8% at Dakar, 40.5% at Cabo Verde), the RMSE are less than 0.03, and the correlations are 80.23% at Dakar station and 82.34% at Cabo Verde. However, for SeaWiFS, the correlation is 13.45% for Dakar and 6.38% for Cabo Verde, and the MRE is 74.54% at Dakar and 60.75% at Cabo Verde. These results are resumed in **Tables 1** and **2**.

4. Conclusion

SOM-NV, an original and efficient method to retrieve optical properties from TOA reflectance measured by satellite-borne multispectral ocean color sensors, has permitted to study realistic spatial distribution of aerosol optical thickness and chlorophyll-a. The method is based on a combination of a neural network classification and a variational optimization. It makes use of the full spectrum of measurements to perform the aerosol identification. The means obtained by SOM-NV appear statistically more representative than those obtained by SeaWiFS. Monthly mean of aerosol optical thicknesses obtained by the standard processing for March and June are drastic because only τ values not exceeding 0.35 are considered. These temporal and spatial failures demonstrate the impossibility of establishing by SeaWiFS global maps of aerosols especially in absorbing environment. The number of pixels processed by SOM-NV even double or triple those computed by SeaWiFS. This is due to the fact that desert aerosols frequently cross the ocean preventing the standard algorithm to retrieve these aerosols and chlorophyll-a below them.

Author details

Daouda Diouf[1]*, Awa Niang[1], Julien Brajard[2], Salam Sawadogo[1], Michel Crepon[2] and Sylvie Thiria[2]

*Address all correspondence to: dadiouf2001@yahoo.fr

1 Ecole Supérieure Polytechnique, Université Cheikh Anta Diop de Dakar, Dakar Fann, Sénégal

2 IPSL/LOCEAN, Université, Paris, France

References

[1] Moulin, C., Gordon, H. R., Banzon, V. F., & Evans, R. H. (2001). Assessment of Saharan dust absorption in the visible from SeaWiFS imagery. Journal of Geophysical Research, 106, 18239–18250.

[2] Diouf, D., Niang, A., Brajard, J., Crepon, M. and Thiria, S. (2013). Retrieving aerosol characteristics and sea-surface chlorophyll from satellite ocean color multi-spectral sensors using a neural-variational method. Remote Sensing of Environment, 130, ISSN: 0034-4257, 74–86. DOI:10.1016/j.rse.2012.11.002.

[3] Shettle, E. P. & Fenn, R. W. (1979). Models of aerosols of the lowers atmosphere and the effect of the humidity variations on their optical properties. Technical Report TR-79-0214, Air Force Geophysical Laboratory, Bedford, Mass.

[4] Kohonen, T. (2001). Self organizing maps (3rd ed.). Berlin Heidelberg: Springer Verlag. 501 pp.

[5] Brajard, J., Jamet, C., Moulin, C. and Thiria, S. (2006). Use of a neuro-variational inversion for retrieving oceanic and atmospheric constituents from satellite ocean colour sensor: application to absorbing aerosols. Neural Networks 19, 178–185.

[6] Jamet, C., Thiria, S., Moulin, C., & Crepon, M. (2005). Use of a neuro-variational inversion for retrieving oceanic and atmospheric constituents from ocean color imagery. A feasibility study. Journal of Atmospheric and Oceanic Technology, 22, (4), 460–475. Doi:10.1175/JTECH1688.1

[7] Niang, A., Badran, F., Moulin, C., Crépon, M. & Thiria, S. (2006). Retrieval of aérosol type and optical thickness over the Mediterranean from SeaWiFS images using an automatic neural classification method. Remote Sensing of Environment, 100, 82–94.

[8] Brajard, J., Niang, A., Sawadogo, S., Fell, F., Santer, R. and Thiria, S.. (2007). Estimating aerosol parameters from MERIS ocean colour sensor observations by using topological maps. International Journal of Remote Sensing, 28, 3–4, 781–795.

[9] Bishop, C. (1995). Neural networks for pattern recognition. Oxford University Press, Inc. New York, NY, USA ISBN:0198538642

[10] Nobileau, D. & Antoine, D. (2005). Detection of blue-absorbing aerosols using near-infrared and visible (ocean color) remote sensing observations. Remote Sensing of Environment, 95, 368–387.

[11] Holben, B., Eck, T., Slutsker, I., Tanre′, D., Buis, J. P., Setzer, E., et al. (1998). AERONET—a federated instrument network and data archive for aerosol characterization. Remote Sensing of Environment, 66, 1–16.

[12] Tanre, D., Deroo, C., Duhaut, P., Herman, M., Morcrette, J., Perbos, J. and Deschamps, P. J. (1997). Description of a computer code to simulate the satellite signal in the solar spectrum: 5s code. International Journal of Remote Sensing, 11, 659–668

Chapter 8

Aerosol–Cloud Interaction: A Case Study

Sandeep R. Varpe, Gajanan R. Aher,

Amol R. Kolhe and Sanjay D. More

Additional information is available at the end of the chapter

http://dx.doi.org/10.5772/65237

Abstract

MODerate Resolution Imaging Spectroradiometer (MODIS) retrieved aerosol and cloud products at the nine selected stations over Western Himalayan and Deccan Plateau regions were inferred to bring out their salient features and to investigate aerosol–cloud interaction. Annually, Ångström exponent (AE) decreases with aerosol optical depth (AOD) while in winter it increases with AOD at most of the stations. Results bring out positive and/or negative association between AOD and almost all the cloud parameters over the selected stations. Aerosol indirect effect (AIE) is quantified for fixed liquid water path (LWP) bins ranging from 1 to 350 g/m^2 at an interval of 25 g/m^2 for three categories of stations, viz., CAT-H, CAT-M, and CAT-L based on heavy, moderate, and low aerosol loading, respectively. AIE is negative at CAT-H (-0.04 ± 0.14), while it is positive at CAT-M (0.01 ± 0.07) and CAT-L (0.10 ± 0.48). During winter, negative AIE has been observed for all three categories. In pre-monsoon, the majority of LWP bins (86% at CAT-H and 60% at CAT-M) showed positive AIE, while about 71% of LWP bins indicted negative AIE at CAT-L. However, during monsoon about 63–71% of LWP bins showed negative AIE at these categories. Study elucidates the influence of factors like cloud type, cloud dynamics/thermodynamics on aerosol–cloud interactions.

Keywords: aerosol–cloud interaction, aerosol indirect effect, aerosol optical depth, Twomey-effect, liquid water path

1. Introduction

Atmospheric aerosols have been found to affect the earth's climate in many characteristic ways [1, 2]. They can affect the energy balance of the earth–atmosphere system by producing a direct or indirect change in the weather and climate system [3]. The direct interaction of aerosols involves both scattering and absorption of radiation, and the relative importance of these processes depends on their chemical composition, refractive index, and size distribution [4, 5]. The indirect effect of aerosols on climate occurs by modifying the cloud optical properties [6]. Thus, the concentration, size, and composition of aerosols which can act as cloud condensation nuclei determine the cloud properties, evolution, and development of precipitation [7]. Aerosols modify cloud properties and precipitation via a variety of mechanisms with varying and contradicting consequences [8].

Cloud interactions with aerosols are hypothesized to be critical to understanding the climate change since clouds play a pivotal role in controlling incoming and outgoing radiation [9]. A large number of studies showed that the anthropogenic aerosols change clouds and their optical properties [4, 10, 11]. Atmospheric aerosols change the concentration and size of the cloud droplets which in turn lead to a change in cloud albedo, its lifetime and thereby affect the precipitation [6, 12]. Also, the reduction in cloud effective radius due to the increase in cloud droplet number concentration (CDNC) leads to the increase in cloud lifetime. The possible repercussion of this process is to decrease the rate of surface evaporation which results in stable and drier atmosphere as a result of the reduction in cloud formation [12]. Anthropogenic aerosols influence mixed-phase clouds in a number of ways and needs comprehensive study to understand the precise phenomenon. A great number of studies were conducted on the possible modification of cloud properties via the interaction with atmospheric aerosol particles, as this may lead to important changes in the Earth's climate. Biomass burning aerosols have been shown to affect clouds through both microphysical and radiative mechanisms [7, 13]. Biomass burning, from both deforestation and annual agricultural burning, is the largest anthropogenic source of such particles in the Southern Hemisphere. Biomass burning aerosols are hygroscopic and can serve as cloud condensation nuclei [14, 15]. More recently, satellite analyses have revealed a persistent correlation between cloud fraction and aerosol optical depth in regions influenced by marine aerosol, smoke, dust, and industrial air pollution [4].

The first indirect effect known as the Twomey effect produces the reduction in cloud effective radius due to the increase in aerosol loading for fixed liquid water path (LWP). Opposite of this effect (i.e., as aerosol loading increases cloud effective radius also increases) were observed over some parts of the world in certain environmental conditions [16]. The Twomey effect and Albrecht effect (i.e., lifetime effect) facilitate cooling of the atmosphere by increasing cloud optical depth (COD) and cloud fraction (CF), respectively [17]. This causes a reduction in the net solar radiation at the top of the atmosphere and hence at the surface. Several other studies have pointed out that the aerosol–cloud interactions are not determined by aerosols alone, but the regional meteorological conditions can play a significant role in this relationship [18]. Comparison of the modeled results and MODerate Resolution Imaging Spectroradiometer (MODIS) retrievals for aerosol indirect effect investigated by Myhre et al. [10] and Storelvmo et al. [18] showed a negative correlation between AOD and cloud effective radius (CER) while a positive correlation was observed between AOD and cloud optical depth (COD).

Extensive studies were conducted on various mechanisms of cloud properties through the interaction of atmospheric aerosol particles with cloud parameters which further influence the

http://dx.doi.org/10.5772/65237

earth's climate. It was found that at low AODs, cloud optical depth (COD) increases with increasing AOD while COD decreases with increasing AOD at higher AODs. This increase was attributed to a combination of microphysical and dynamical effects, while the decrease was due to the dominance of radiative effects that thin and darken clouds [19]. The AOD and cloud fraction correlation increases for those regions which have more particulate matter due to dust, biomass, industrial, and domestic activities [20].

In the present study, we have analyzed 11 years of MODIS aerosol and cloud products at the selected stations over Western Himalayan (75°E–80°E; 29°N–33°N) and Deccan Plateau (73°E–75°E; 16°N–19°N) regions. The selected stations over these regions comprised of (1) Dharmashala (DSL), (2) Mandi (MND), (3) Shimla (SML), (4) Ludhiana (LDN), (5) Patiala (PTL), (6) Muzaffarnagar (MZR), (7) Pune (PUN), (8) Satara (STR), and (9) Kolhapur (KPR) (abbreviated hereinafter as DSL, MND, SML, LDN, PTL, MZR PUN, STR, and KPR, respectively). Western Himalaya is desert dust dominated region while Deccan Plateau is dominated by fossil fuel and biomass burning. The study deals with the spatiotemporal variations of the MODIS retrieved aerosol and cloud products at the nine selected stations over Western Himalayan and Deccan Plateau regions to bring out their salient features. The data were also employed to investigate aerosol–cloud interaction and to quantify the aerosol indirect effect (AIE) over these regions.

2. Topography of study region

The Deccan Plateau is a large region that covers most of South Central India. The average elevation is 1000–2000 feet (305 and 610 m) above sea level along the northern sections of the region and 2000–3000 feet (610 and 915 m) in the southern section. Red loam or sandy loam soil usually overlies the granites and metamorphic rocks resulting in less fertile and less moisture retentive soil than found in the North Deccan region. The region slopes generally eastward allowing the drainage to flow toward the Bay of Bengal. Water in the rivers fluctuates considerably during the monsoon and the dry seasons. Their only source of water is the monsoon rains unlike rivers flowing out of the Himalayas that have year-round moisture from snow packs in the high mountains. In the winter or dry season, many of the rivers throughout the South Deccan become almost dry and are useless for irrigation. Also, some of the rivers flow through well-incised valleys, allowing little space for a flood plain and making it nearly impossible to direct water for irrigation onto the adjoining uplands. The climate is generally semi-arid with <35 inches (89 cm) of rainfall. Ironically, the Western Ghats are only 30–40 miles (48–64 km) away with annual precipitations exceeding 100 inches (254 cm).

In the Indian part of the Western Himalayas the surface weather elements, like precipitation and temperature, are intensely governed by local topography and local atmospheric circulations. The different altitude and orientation of the Himalayan ranges give rise to different thermo dynamical and dynamical forcing. Topography, heterogeneity, and land use variability are the characteristics of Western Himalayas (WH). Western disturbances (WDs), embedded in large-scale westerlies are responsible for winter precipitation, mainly snow, in WH. The interplay of topography with WDs determines orographic precipitation over the Himalayan region.

3. Data products and methodology

Satellite observations have advantages over the ground-based measurements, in that, they provide information over the larger spatial domain [21]. The MODIS was designed with aerosol and cloud remote sensing in mind [22]. The MODerate Resolution Imaging Spectroradiometer (MODIS) aboard the Terra (launched in 1999) and Aqua (launched in 2002) monitors the earth–atmosphere system twice daily over a given location. It is sun-synchronous and near polar orbiting satellite with a circular orbit of 705 km above the surface. MODIS has 36 bands ranging from 0.4 to 14.4 μm wavelengths with three different spatial resolutions (250, 500, and 1000 m) and views the Earth with a swath of 2330 km, thereby providing near-global coverage on daily basis, with equatorial crossing local time of 10:30 am and 1:30 pm for Terra and Aqua, respectively (http://modis.gsfc.nasa.gov/).

Among the hundreds of products derived from MODIS-measured radiances are a suite of aerosol products [23] and another set of cloud products [24], including aerosol optical depth (AOD), cloud top pressure, and cloud fraction. Often, the AOD is used as a proxy for the cloud condensation nucleus (CCN) concentration. The reliability of this proxy depends on the uniformity of the aerosol size, composition, vertical distribution, but may in many cases be used as a first approximation. The MODIS data are available at different processing levels, level 1.0 (geolocated radiance and brightness temperature), level 2.0 (retrieved geophysical data products) and level 3.0 (gridded points) [24]. MODIS uses infrared bands to determine the physical properties of cloud in relation to cloud top pressure and temperature, and visible and near-infrared bands to determine optical and microphysical cloud properties [23, 25]. For water vapor, the retrieval the near-infrared region is adopted.

For this study, simultaneously retrieved datasets from MODIS (Terra) for the period 2000–2010 were used. Terra MODIS level_3 (C_005) monthly data products of aerosol optical depth (AOD), Ångström exponent (AE), cloud fraction (CF, day), water vapor (WV, above cloud), cloud effective radius (CER, liquid), liquid water path (LWP), and cloud optical depth (COD, liquid) were retrieved over the study region. As shown in **Figure 1**, the study regions were divided into 1° × 1° grid box centered at (1) DSL (32°16′ N; 76°23′E), (2) MND (31°43′ N; 76°58′ E), (3) SML (31°06′ N; 77°13′ E), (4) LDN (30°55′ N; 75°54′E), (5) PTL (30°20′ N; 76°25′ E), (6) MZR (29°28′ N; 77°44′E), (7) PUN (18°31′ N; 73°55′ E), (8) STR, (17°42′ N; 74°02′ E), and (9) KPR(16°42′ N; 74°16′ E).

For the estimation of aerosol indirect effect (AIE) on the basis of the observed AODs, we grouped the selected stations into different categories viz., CAT-H (heavy aerosol loading), CAT-M (moderate aerosol loading), and CAT-L (low aerosol loading). The MODIS (Terra) level_3 (C_005) daily data products of AOD, LWP, and CER were retrieved for this estimation, and analysis was performed over each category and as well as over each station keeping the fixed LWP constraint. The retrieved LWP and CER were divided into 14 different bin sizes.

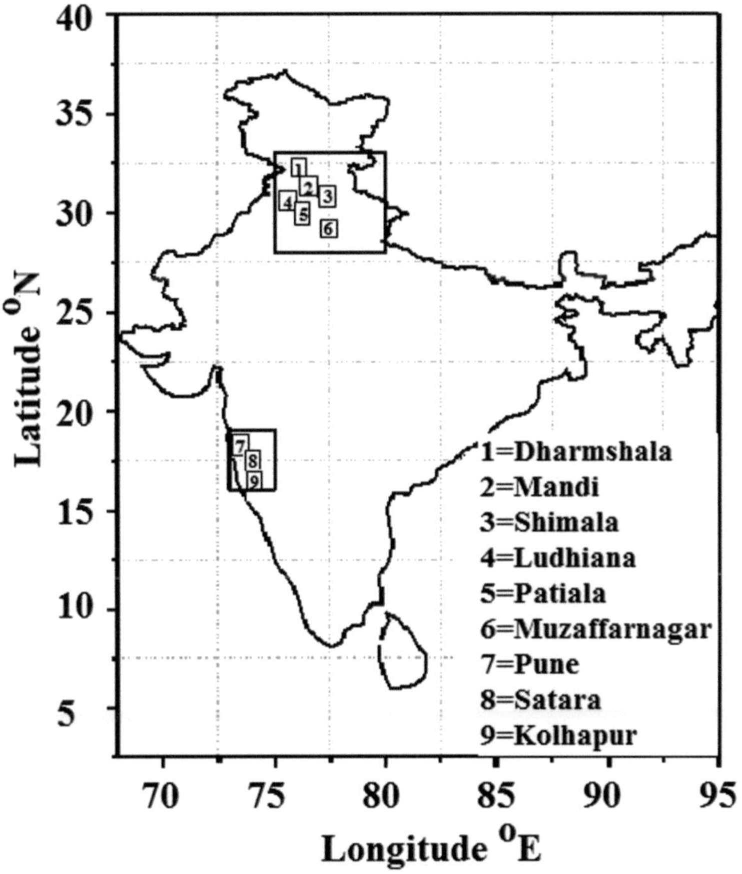

Figure 1. Study regions selected based on the dominant aerosol sources over India.

AIE was estimated for different seasons and for the entire study period 2000–2010 by evolving a linear least square fit to the plot between CER and AOD at fixed LWP and using the following equation [26],

$$\mathrm{AIE} = -\frac{d \ln r_e}{d \ln \tau_a} \tag{1}$$

Here, r_e is the cloud effective radius (CER) for fixed LWP and τ_a is the AOD. The degree of significance of AIE and correlation coefficients of linear regression fit has been also determined over the selected stations using two-tailed t-tests at 90 and 95% of confidence level. The correlation coefficients between AOD and other parameters (AE, CF, COD, CER, LWP, and WV) for 11 years data at each station are given in **Table 2**, and seasonal correlations of these parameters are given in **Table 3**. In these tables, the doubly underlined correlation coefficients are significant at 0.05 level (95% confidence level) while singly underlined correlation coefficients are significant at 0.1 level (90% confidence level) and the rest are less significant.

4. Results and discussion

4.1. Spatial climatology of aerosol optical depth

The spatial distribution of the monthly mean AODs for the period 2000–2010 at a wavelength 550 nm has been derived and is plotted in **Figure 2**. It reveals that the aerosol loading (i.e., AOD) have considerable influence on the atmosphere over Western Himalayan, Deccan Plateau regions, and their associated stations viz., DSL, MND, SML, LDN, PTL, MZR, PUN, STR, and KPR. The spatial distribution of AOD depicts occurrence of consistently high AODs (~0.7) over the Western Himalaya. It was found that the long-range transport of desert dust from the Arabian Peninsula contributes to the net regional aerosol loading with the marked increase in AOD over the Himalayan region in pre-monsoon months [27]. Over the Himalayan region, dust transport reaches up the slopes of the Himalayas and is further vertically elevated to higher altitudes because of the strong westerly pre-monsoon winds coupled with enhanced convection and pressure gradient resulting from large topographic differences [27]. High AODs have also been observed over the Ganga basin, situated in the Northern part of India [28]. The prevalence of high AODs over the Northern India is attributed to dense population density, the presence of heavy industries in the region, and the transport of desert dust from the nearby the Thar Desert in Rajasthan [29–31]. The spatial gradient of AOD shows an increase from the southern part of the Indian subcontinent to northern part up to the Himalayas [28]. Comparatively, AOD values over Deccan Plateau region (~0.35 to 0.4) are lower than WH region which can attribute to the anthropogenic activities, like the industrial region [32].

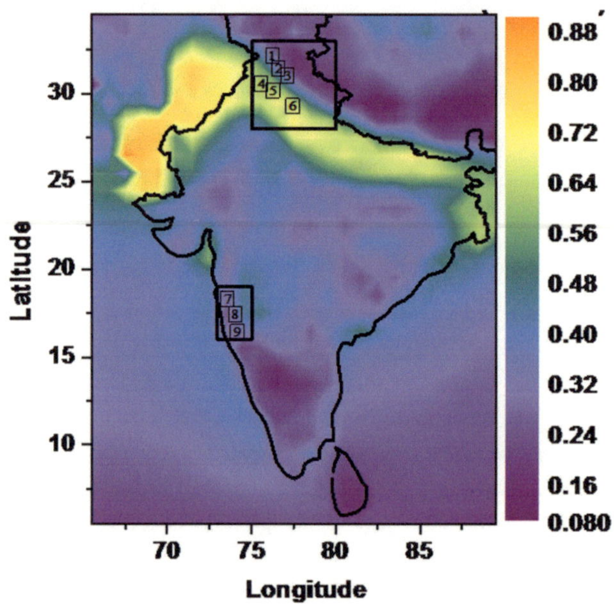

Figure 2. Spatial variation of monthly mean $AOD_{550\,nm}$ during 2000–2010.

http://dx.doi.org/10.5772/65237

4.2. Seasonal variation of AOD

The mean seasonal variation of $AOD_{550\ nm}$ wavelength was derived by considering MODIS (Terra) retrieved AODs over Western Himalayan and Deccan Plateau regions (i.e., stations associated with these regions) for the period of study. The resulting mean seasonal variation of AOD, for the period 2000–2010, elucidated in **Figure 3**, reveals a statistically significant seasonal variation of AODs during monsoon and other seasons. The seasonal mean AODs, given in **Table 1**, show that AOD was lowest during winter (0.29 ± 0.05) and highest during monsoon (0.60 ± 0.25) followed by pre-monsoon (0.42 ± 0.10) at all the stations. Higher AODs during the monsoon may be because of the hygroscopic growth of water-soluble aerosols and transport of larger sized aerosols (dust and sea salt) during favorable wind conditions [29]. Also, the ensemble mean monsoon AOD may be high due to limitations on the numbers of data points which were very few, because of the prevalence of overcast conditions (cloudy in nature) during most of the days. The similar increase in AOD during pre-monsoon and monsoon season has previously been reported for Indian Subcontinent [28].

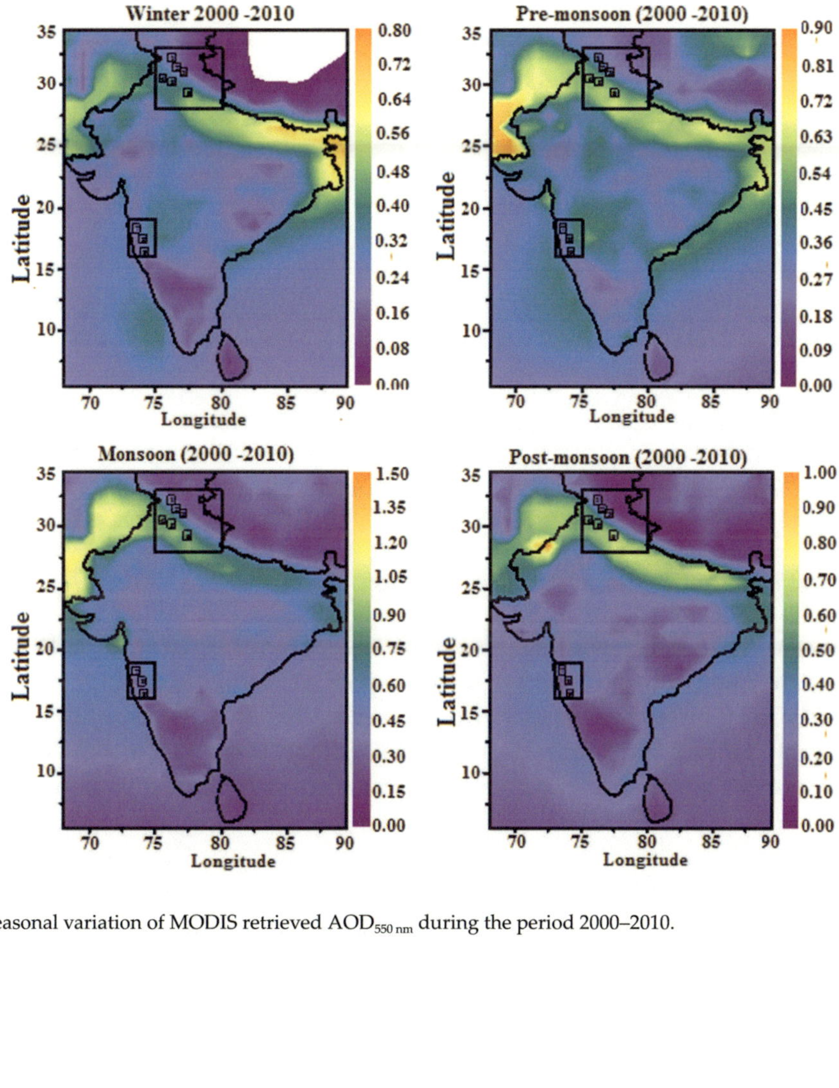

Figure 3. Seasonal variation of MODIS retrieved $AOD_{550\ nm}$ during the period 2000–2010.

Station wise, LDN, PTL, and MZR in Western Himalayan region depict higher mean AODs over the period 2000–2010 as compared to the other investigated cities viz., PUN, STR, and KPR in Deccan Plateau region. AODs at LDN, PTL, and MZR in pre-monsoon were found to be 0.53, 0.51, and 0.54, respectively, whereas the corresponding winter AODs at these places were found to be 0.35, 0.34, and 0.45, respectively. On the other hand, over PUN, STR, and KPR stations, AODs were found to be 0.30, 0.36, and 0.32, respectively, in pre-monsoon and 0.24, 0.26, and 0.23, respectively, during winter. Analysis indicates that the AODs generally increase from lower latitude to higher latitude [32]. The occurrence of high AODs at LDN, PTL, and MZR is due to their close proximity to the Thar Desert where frequent dust storms occur during pre-monsoon months. In addition, the presence of higher temperatures tends to hold more water vapor favoring aerosols to grow in size causing higher AODs at these sites. Aerosol loading over the studied region is low after the monsoon season as rainfall washes out most of the aerosol concentration [33]. The post-monsoon season shows a continuation of the trend observed for AOD distribution in winter.

Stations	Seasonal mean AOD and standard deviation			
	Winter	Pre-monsoon	Monsoon	Post-monsoon
Dharmashala	0.25 (±0.04)	0.37 (±0.11)	0.51 (±0.19)	0.31 (±0.07)
Mandi	0.22 (±0.04)	0.36 (±0.13)	0.49 (±0.20)	0.26 (±0.07)
Shimla	0.23 (±0.03)	0.37 (±0.11)	0.51 (±0.20)	0.28 (±0.07)
Ludhiana	0.35 (±0.05)	0.53 (±0.17)	0.86 (±0.37)	0.55 (±0.14)
Patiala	0.34 (±0.06)	0.51 (±0.16)	0.81 (±0.34)	0.51 (±0.13)
Muzaffarnagar	0.45 (±0.07)	0.54 (±0.17)	0.82 (±0.30)	0.70 (±0.14)
Pune	0.24 (±0.05)	0.30 (±0.05)	0.43 (±0.23)	0.24 (±0.06)
Satara	0.26 (±0.06)	0.36 (±0.05)	0.51 (±0.24)	0.25 (±0.05)
Kolhapur	0.23 (±0.06)	0.32 (±0.06)	0.46 (±0.20)	0.25 (±0.05)

Table 1. MODIS (Terra) derived mean seasonal AODs at the selected stations over Western Himalaya and Deccan Plateau Region during 2000–2010.

4.3. AOD and Ångström exponent (AE) correlation

Figure 4 shows the monthly mean composite plots of AOD and AE (α) for the selected stations in the Western Himalayan and Deccan Plateau regions during 2000–2010. Also, **Table 2** displays the correlation coefficients of linear regression analysis between AOD and AEs. It is clear from both **Table 2** and **Figure 4** that the MODIS retrieved AODs and AEs are either inversely or positively correlated and the correlations are significant at 95% confidence level at all the locations. It can be noticed that during monsoon season, AE is negatively correlated with AOD with 95% confidence level while in winter it shows positive correlations with AOD at all the stations. The positive correlation of AE with AOD was observed at the locations STR and KPR during pre-monsoon and at MZR, PUN, and KPR during post-monsoon season. This

http://dx.doi.org/10.5772/65237

kind of behavior may be attributed to the hygroscopic growth of the aerosol particles producing a significant shift in aerosol size spectrum which in turn can substantially influence the magnitude of the Ångström exponent. More particularly, the change in the ratio of small to large particles can also bring about a change in the Ångström exponent. A dry particle of anthropogenic origin may have a Ångström exponent which is 60% higher than a particle with a growth factor of 1.6, which illustrates that hygroscopic growth, can substantially impact the Ångström exponent [10]. It may be possible that high AOD during pre-monsoon is due to the increase in the size of the humidified aerosols present near the cloudy area.

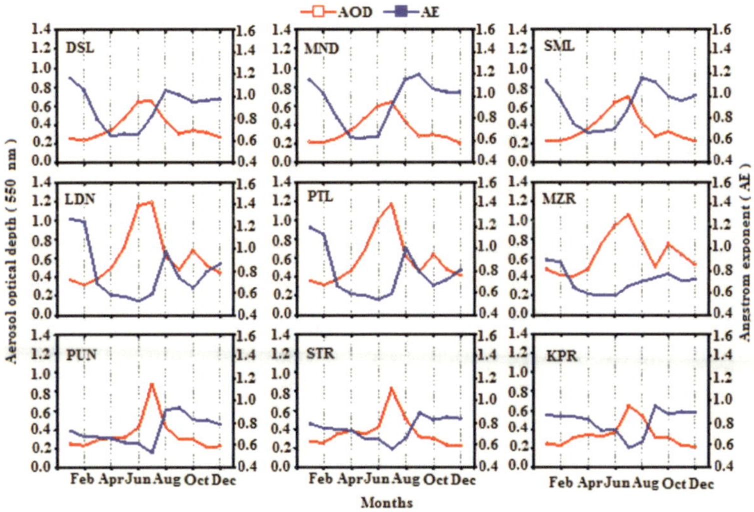

Figure 4. Monthly mean composite plots of AOD and AE over the selected stations in the Western Himalayan and Deccan Plateau regions during 2000–2010.

Stations	AE	CF	COD	CER	CWP	WV
DSL	**−0.48**	**0.53**	−0.07	**0.19**	0.09	**0.74**
MND	**−0.43**	**0.57**	0.12	**0.22**	**0.23**	**0.75**
SML	**−0.39**	**0.57**	0.02	**0.24**	*0.17*	**0.74**
LDN	**−0.54**	**0.60**	0.05	**0.19**	*0.16*	**0.66**
PTL	**−0.50**	**0.61**	0.03	**0.23**	0.14	**0.71**
MZR	**−0.31**	**0.54**	−0.03	**0.23**	0.09	**0.67**
PUN	**−0.26**	**0.57**	**0.34**	**0.19**	**0.43**	**0.37**
STR	**−0.60**	**0.60**	**0.40**	**0.33**	**0.48**	**0.37**
KPR	**−0.52**	**0.60**	**0.43**	**0.23**	**0.49**	**0.38**

Correlation coefficients bolded are significant at 0.05 level (95% confidence level), while italicized are significant at 0.1 level (90% confidence level) and the rest are less significant.

Table 2. Correlation coefficients of the linear regression analysis between AOD against AE and cloud parameters at each station for the period of 2000–2010.

The Ångström exponent is an approximate measure of the aerosol particle size distribution; smaller the Ångström exponent larger is the size of the particles [34]. **Figure 4** reveals that there is a decrease in Ångström exponent values with increase in the AODs at all sites. Higher values of α indicate a sharper aerosol size spectrum. It clearly shows the dramatic transformation in the aerosol spectra, from high accumulation mode domination during September through February months (revealed by values of $\alpha > 1.0$ and steep AOD spectra) to increased coarse mode domination ($\alpha < 1.0$) during March through June–July months. Lower values of AE clearly indicate that the dominance of coarse-mode aerosols which affirms the influx of transported coarse-mode dust aerosols from the Arabian Desert and the Thar Desert [31, 35].

4.4. Interaction between AOD and cloud parameters

The MODIS provides an enormous amount of data which is valuable for understanding how aerosols influence clouds [10]. Aerosols are highly variable in space and time with different properties, characteristics, and concentrations. It is this peculiar nature of aerosols along with prevailing the meteorological conditions that brings in a complicating factor and makes the aerosol–cloud interaction a complex phenomenon. Rain formation depends on the number concentration of aerosols in atmospheric pollution and is unaffected by low pollution; however, during heavy pollution episode, rainfall is significantly reduced. Aerosols act as ice nuclei and cloud condensation nuclei. The presence of small condensation nuclei in the atmosphere can initiate the suppression of heterogeneous freezing with the result that many small droplets remain liquid below the homogeneous freezing temperature [36]. However, most of the observational studies are based on specified cases from which it is difficult to determine as to which effects are more significant and dominant and thus their long-term implications remain unknown [37].

In the present work, an attempt has been made to understand and quantify the influential role of AOD on cloud parameters by analyzing correlations amongst them on spatial scale and by estimating AIE for the nine selected stations. The results are highlighted in the following subsections through the use of spatial correlation. In addition, we have estimated the seasonal correlation coefficients for each parameter at the selected stations as well as for the entire period of 2000–2010. The aerosol–cloud relationship has also been studied using the monthly mean composite plots for 11 years of data so used.

4.4.1. Influence of AOD on cloud fraction

The co-variation plot of $AOD_{550\ nm}$ and CF during the period 2000–2010 has been constructed over the selected nine sites in the Western Himalayan and Deccan Plateau region in India, and the results are displayed in **Figure 5**. Also, linear regression correlation coefficients for the plots of $AOD_{550\ nm}$ against AE and MODIS-retrieved cloud parameters viz., CF, COD, CER, CWP, and WV have been estimated for the selected time period of 11 years on a seasonal basis. The resulting data are shown in **Table 3** which reveals that the satellite retrieved cloud fraction data shows a strong positive correlation with $AOD_{550\ nm}$ at all the selected locations in the respective study regions. The present results are found to be consistent with those reported by Myhre et al. [10] and Kaufman et al. [4]. Interestingly, the correlation coefficients at heavy columnar aerosol loading locations (AOD ranging between 0.59 and 0.65), i.e., LDN, PTL, and MZR and at relatively low columnar aerosol loading locations (AOD ranging between 0.31 and 0.36), i.e., PUN, STR, and KPR are more or less similar. It is important to note here that the Western Himalayan region is desert-dust dominated region due to its proximity to Thar Desert, whereas Deccan Plateau region is dominated by anthropogenic activity in the form of industrial/vehicular pollution and biomass burning. Consequently, the occurrence of higher correlations between $AOD_{550\ nm}$ and CF in these anthropogenically, biomass burning and desert-dust dominated aerosol regions indicate that the meteorological factors significantly influence this relationship [32]. Recently, it is found that the increasing aerosol concentration leads to increase in cloud cover and showing that the aerosol concentrations change the cloud properties [38]. This is because regions of low atmospheric pressure have more tendency to create conditions necessary for cloud formation by accumulating aerosol particles and water vapor [39].

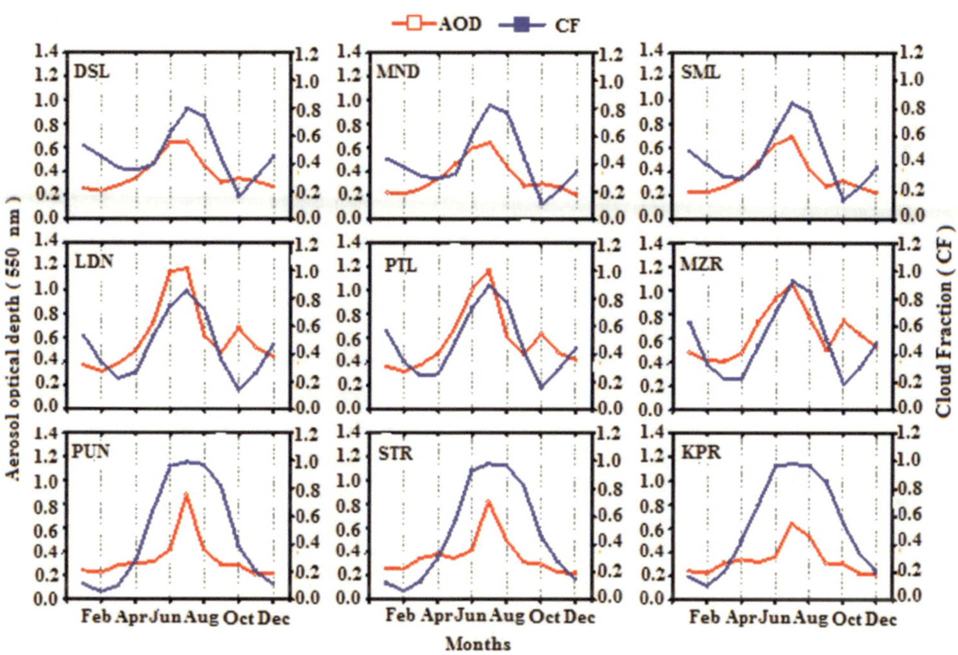

Figure 5. Monthly mean composite plots of AOD and CF at the selected stations in the Western Himalayan and Deccan Plateau regions during 2000–2010.

The monthly mean composite plots of AOD and CF during the period 2000–2010 (**Figure 5**) show that the CF increased with AOD at all locations throughout the entire period of study. The CF and AOD variations demonstrate an out of phase correlation during the winter season for the Western Himalayan stations. Also, the relationship becomes negative in the months of July and August as CF increased with a decrease in AOD. The magnitudes of the seasonal correlation coefficients between AOD and CF, given in **Table 3**, signify that the CF increases with AOD at all the station (except SML; negative correlations) during winter (DSL = 0.24,

MND = 0.14, LDN = 0.65, PTL = 0.47, MZR = 0.58, PUN = 0.42, STR = 0.65, KPR = 0.74), pre-monsoon (DSL = 0.31, MND = 0.34, LDN = 0.77, PTL = 0.74, MZR = 0.79, PUN = 0.30, STR = 0.10, KPR = 0.20) and monsoon(DSL = 0.51, MND = 0.49, LDN = 0.63, PTL = 0.61, MZR = 0.48, PUN = 0.46, STR = 0.53, KPR = 0.41) seasons. Myhre et al. [10] have found that the decrease of CF with increasing AOD occurs mainly for low AODs (below ~0.3) and also reported an increase in AOD with an increase in RH due to the swelling of hygroscopic particles which are near to the clouds. The cloud cover also exhibits a weak negative correlation with the potential temperature lapse rate, and vertical shear of the horizontal wind in the middle atmosphere [40]. This gives rise to shift in aerosol size distribution with smaller particles coming into optical range while larger particles moving out of it [41]. Positive correlations between CF and AOD appear to be the reflection of this effect. This increase may be due to the complexity of the domain, the type of land surface (albedo), the choice (classification) of aerosol mixtures applied in the MODIS retrieval for that particular area, the impact of meteorology on aerosol transport, and the aerosol chemistry. Kaufman et al. [4] mentioned that the cloud cover increases with increase in aerosol concentrations and found that the cloud properties change as a result of variations in large-scale atmospheric circulation and also affect aerosol concentrations. For example, regions of low atmospheric pressure are convergence zones that tend to accumulate aerosols and water vapor, thus generating conditions favorable for cloud formation [42].

Station	Parameters	Winter	Pre-monsoon	Monsoon	Post-monsoon
DSL	AE	0.215	**−0.580**	**−0.687**	−0.136
	CF	0.248	*0.316*	**0.519**	−0.184
	COD	0.285	**−0.344**	−0.058	−0.228
	CER	−0.184	−0.243	**−0.367**	0.317
	LWP	0.187	**−0.452**	−0.233	*−0.071*
	WV	0.243	**0.795**	**0.573**	**0.399**
MND	AE	0.106	**−0.589**	**−0.617**	−0.179
	CF	0.138	**0.345**	**0.494**	−0.274
	COD	0.295	−0.156	−0.027	−0.283
	CER	0.078	−0.140	−0.221	*0.330*
	LWP	0.259	−0.266	−0.141	−0.128
	WV	0.070	**0.774**	**0.598**	**0.466**
SML	AE	0.057	**−0.571**	**−0.659**	−0.192
	CF	−0.044	*−0.299*	**−0.342**	**0.406**
	COD	0.114	−0.157	−0.159	−0.212
	CER	−0.044	*−0.299*	**−0.342**	**0.406**
	LWP	0.116	−0.265	**−0.325**	0.004
	WV	0.274	**0.791**	**0.507**	**0.515**
LDN	AE	0.062	**−0.796**	**−0.607**	**−0.483**
	CF	**0.659**	**0.779**	**0.636**	**−0.361**
	COD	0.177	−0.057	0.205	−0.221
	CER	**−0.451**	−0.027	−0.122	0.220
	LWP	−0.077	−0.037	0.0596	−0.087
	WV	**0.577**	**0.855**	*0.262*	**0.445**
PTL	AE	0.122	**−0.750**	**−0.645**	**−0.352**

Station	Parameters	Winter	Pre-monsoon	Monsoon	Post-monsoon
	CF	**0.471**	**0.747**	**0.614**	**−0.371**
	COD	0.356	−0.104	−0.158	0.085
	CER	**−0.446**	−0.144	−0.238	**0.523**
	LWP	0.144	−0.123	**−0.331**	0.262
	WV	**0.596**	**0.848**	**0.449**	**0.378**
MZR	AE	0.172	**−0.686**	**−0.436**	**0.438**
	CF	**0.575**	**0.795**	**0.484**	−0.243
	COD	0.352	−0.108	−0.121	0.0099
	CER	**−0.634**	−0.280	−0.169	**0.392**
	LWP	−0.025	−0.168	−0.237	0.220
	WV	*0.383*	**0.851**	**0.493**	**0.481**
PUN	AE	**0.674**	−0.206	**−0.478**	0.165
	CF	*0.428*	*0.306*	**0.469**	**0.409**
	COD	−0.210	0.124	0.222	0.131
	CER	0.180	**−0.349**	**0.518**	−0.054
	LWP	−0.096	0.007	**0.320**	0.074
	WV	0.097	0.171	**−0.432**	*0.340*
STR	AE	**0.694**	0.113	**−0.628**	−0.136
	CF	**0.656**	0.100	**0.533**	**0.434**
	COD	−0.055	0.004	**0.364**	0.173
	CER	−0.323	−0.181	**0.400**	0.083
	LWP	−0.143	−0.132	**0.458**	0.195
	WV	0.127	−0.090	**−0.565**	*0.426*
KPR	AE	**0.704**	0.128	**−0.575**	0.049
	CF	**0.742**	0.206	**0.418**	**0.583**
	COD	−0.052	0.026	**0.389**	0.235
	CER	−0.260	−0.027	0.185	−0.105
	LWP	−0.159	−0.055	**0.430**	0.272
	WV	0.322	0.017	**−0.303**	*0.465*

Correlation coefficients bolded are significant at 0.05 level (95% confidence level), while italicized are significant at 0.1 level (90% confidence level) and the rest are less significant.

Table 3. Same as in **Table 2** but for different seasons during the period 2000–2010.

4.4.2. Association between AOD and cloud effective radius

The cloud effective radius (CER) is the weighted mean of the size distribution of cloud droplets in the atmosphere. CER (defined as the ratio of the third to the second moment of a droplet size distribution) is one of the key variables that are used for calculation of the radiative properties of liquid water clouds [43]. Analysis of the present data reveals that the CER and AOD values are positively correlated at all the locations during the study period 2000–2010 (**Table 2**). These findings are found to be consistent with the results of Yuan et al. [44] which manifested that the direct correlation between AOD and CER may be due to various artifacts as well as aerosol swelling, partial cloudiness, atmospheric dynamics, cloud three-dimensional (3D), and surface influence effects. Hygroscopic aerosols grow in size as a consequence of moisture uptake from water vapor. The size and the refractive index of the aerosol change due to the hygroscopic growth of the aerosols resulting in the subsequent increase in AOD [26]. For example, aerosols which are in close proximity to the cloud swell more because of the higher moisture content present near the cloud producing larger AOD. Simultaneously, deeper clouds tend to have larger droplets than shallower clouds.

On the other hand, **Figure 6** shows that CER values are found to be lower in pre-monsoon months (0.4–0.6) and higher in monsoon months (0.8–1.3). Results clearly indicate that the Western Himalayan and Deccan Plateau regions receive maximum rainfall during the monsoon months. Bhawar and Devara [45] carried out the similar study over Pune and found that increase in CER produces an increase in COD which could result in more rainfall and *vice-versa*. The present results, on the seasonal scenario (shown in **Table 3**), also reveal that the CER is negatively correlated with AOD during winter and post-monsoon seasons at most of the stations. Analysis of the MODIS retrievals indicates that the negative/positive correlation of CER with AOD is not determined by aerosol–cloud interaction alone. Instead, the regional meteorological conditions, as well as aerosol type, clouds dynamics, and thermodynamics of the atmosphere, can play the significant role in this relationship.

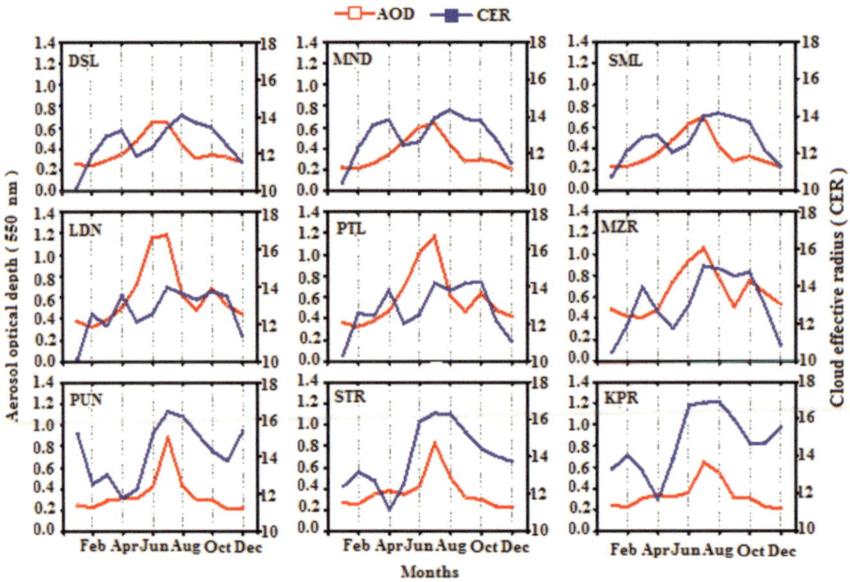

Figure 6. Monthly mean composite plots of AOD and CER at the selected stations in the Western Himalayan and Deccan Plateau regions during 2000–2010.

4.4.3. Effect of AOD on liquid water path

Results displayed in **Table 2** for the correlations between AOD and cloud parameters depict occurrence of relatively high positive correlations between AOD and LWP at PUN, STR, and KPR [correlation coefficient (R) ranging between 0.43 and 0.49] than that at DSL, MND, SML,

LDN, PTL, and MZR [correlation coefficient (R) ranging between 0.09 and 0.23] during 2000–2010. This indicates that the increase of LWP with AOD is consistent with the previous studies of Yuan et al. [44]. It suggests that the assumption, i.e., the first indirect effect is not always valid [44]. However, it should be noted that the MODIS LWP is not an independent measurement but estimated from the respective values of COD and CER.

Monthly variation of AOD and LWP for the study period (**Figure 7**) and their seasonal correlation analysis (**Table 3**) points out that the sign of correlation (negative/positive) changes from season to season and from place to place indicating spatiotemporal variability in these correlations. In winter, LWP decreases with increasing AOD at the stations LDN, MZR, PUN, STR, and KPR while it increases with AOD at DSL, MND, SML, and PTL. In pre-monsoon, at all the stations (except PUN), LWP was found to decrease as AOD increased. During monsoon, LWP decreases with AOD at stations DSL, MND, SML, PTL, and MZR while it increases with AOD at LDN, PUN, STR, and KPR. For the post-monsoon season, at the stations DSL, MND, and LDN, a negative relationship was observed between AOD and LWP while it was positive at the stations PTL, MZR, PUN, STR, and KPR. The possible reason behind the reduction of LWP during winter, pre-monsoon, monsoon, and post-monsoon seasons may be warming of clouds due to the dust aerosols, absorbing incoming solar radiation thereby increasing evaporation of cloud droplets leading to the reduction of LWP, i.e., the so-called semi-direct effect [46].

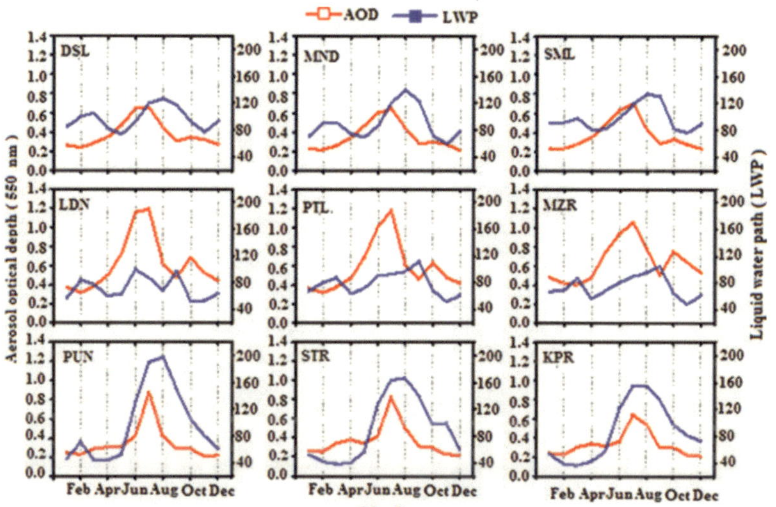

Figure 7. Monthly mean composite plots of AOD and LWP at the selected stations in the Western Himalayan and Deccan Plateau regions during 2000–2010.

Statistical analysis shows that during 2000–2010, about 45 and 11% of the correlation coefficients at the selected stations were significant at 95 and 90%, respectively. During winter, about 56% of the correlation coefficients at the selected stations were consistent with the existing hypothesis (i.e., the negative relationship between AOD and LWP). At the majority of the stations, during pre-monsoon, negative correlations between AOD and LWP were found to be prevalent for about 89% of the cases, out of which 13% were significant at 95% level. In monsoon, about 56% of the correlation coefficients showed an inverse relationship between AOD and CER; of these about 40% correlation coefficients were significant at 95% confidence level. In all, about 67% of correlation coefficients displayed inconsistency with the Twomey effect and about 33% of the correlation coefficients showed consistency with Twomey effect out of which about 33% of the correlation coefficients were significant at 90% level.

4.4.4. Correlation between AOD and water vapor

Atmospheric water vapor content is an important component of the global climate system and plays an important role in the radiation balance of the earth–atmosphere system [47]. It is the basic and crucial component for the cloud formation. Investigation of the possible changes in column water vapor in association with aerosols may be of considerable importance to give insight into the aerosol impact and speed of hydrological cycle [10]. MODIS retrievals provide column water vapor (CWV) in the clear sky and cloudy or above clouds individually. In the present study, we have used CWV above cloud data retrieved from Terra for the period of 2000–2010. The correlation coefficients were determined at each station situated in both Western Himalayan and Deccan Plateau regions during the period 2000–2010. Analysis (shown in **Table 2**) reveals that at higher latitudes, AOD, and WV have strong positive correlation than at lower latitudes. Strong positive correlations were found over DSL (0.74), MND (0.75), SML (0.74), LDN (0.66), PTL (0.71), and MZR (0.67), whereas relatively low correlation was observed over PUN (0.37), STR (0.37), and KPR (0.38).

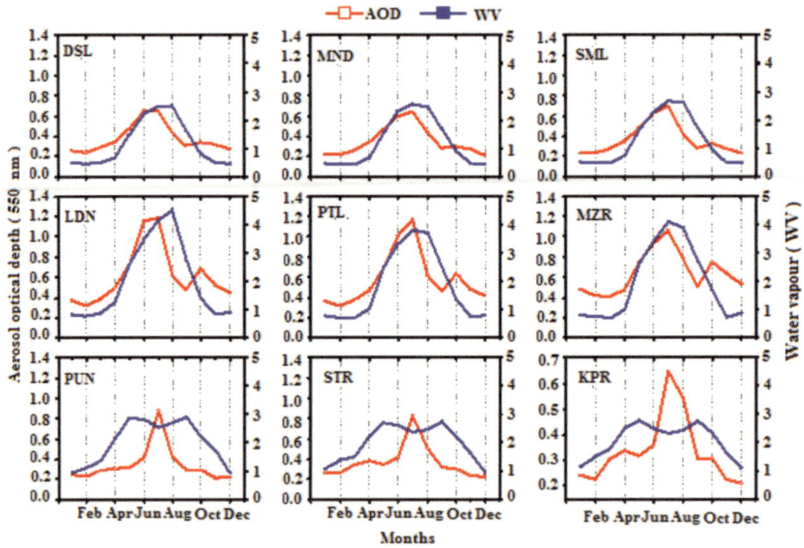

Figure 8. Monthly mean the composite plot of AOD and WV at the selected stations in the Western Himalayan and Deccan Plateau regions during 2000–2010.

http://dx.doi.org/10.5772/65237

The seasonal correlations between AOD and WV (**Table 3**) reveal that AOD and WV have a strong positive correlation (>0.6) over Western Himalayan region and moderate correlation (0.4–0.6) exist at Deccan Plateau region during the pre-monsoon, monsoon, and post-monsoon seasons. There was no significant correlation (<0.32) between AOD and WV during the winter season at all the stations, since the dust aerosols were less common or even absent during winter, consequently, less WV was observed.

Low AOD values in winter, a global feature, is attributed to the removal of aerosols due to monsoon rains and decreased aerosol input due to a colder ground surface. Additionally, the less hygroscopic growth of aerosols due to low WV content may lead to relatively lower AOD values [48]. Also, the decrease in WV with AOD may be a consequence of the increased aerosol loading, since more particles act as cloud condensation nuclei (CCN) followed by an increase in cloud droplet number concentration. Moreover, hygroscopic particles uptake more moisture and this altogether leads to a decrease in WV. It is interesting to note that at the stations PUN, STR, and KPR an abrupt increase in LWP (see Section 4.4.3) was observed during JUN to AUG leading to rapid increase in CER (see Section 4.4.2) and hence the COD. This also supports our hypothesis that WV must have been utilized for the droplet formation thereby decreasing WV. Statistical analysis reveals that at all the selected station's correlation coefficients between AOD and WV during 2000–2010 showed positive AOD-WV relationship, and all the correlation coefficients were significant at 95% level. Seasonally, about 89% of the correlation coefficients at the selected stations evidenced positive correlations in all the seasons. Out of these, about 66 and 16% of correlation coefficients are significant at 95 and 90% confidence level, respectively.

The relation between aerosol and WV has an implication for the radiative forcing both through direct and indirect mechanisms [9]. EL-Askary and Kafatos [49] have found that aerosols cause a reduction in cloud droplet size and hence lead to suppression of the precipitation. The direct effect results in radiation scattering due to an increase in aerosol particle size, accompanied by the uptake of WV. The black cloud episode is comprised mainly of anthropogenic pollutants acting as cloud condensation nuclei leading to the formation of WV cover. The present analysis demonstrates the presence of high WV over locations where high AOD is observed (**Figure 8**).

4.5. Estimation of aerosol indirect effect

The aerosol indirect effect has been estimated using Eq. (1) under fixed LWP as the constraint. For this, magnitudes of LWP are divided into 14 different bins at an interval of 25 g/m^2 viz. 1–25, 25–50, 51–75, 76–100, 101–125, 126–150, 151–175, 176–200, 201–225, 225–250, 251–275, 276–300, 301–325 and 326–350 g/m^2 each. Aerosol indirect effects was estimated seasonally as well as for the entire study period 2000–2010 at each category, and the results are shown in **Figures 9** and **10**, respectively. The two-tailed t-test has also been carried out for determination of the significance of aerosol indirect effect at 90 and 95% confidence levels which are indicated by gray color solid bar and black color solid bar in these figures, respectively.

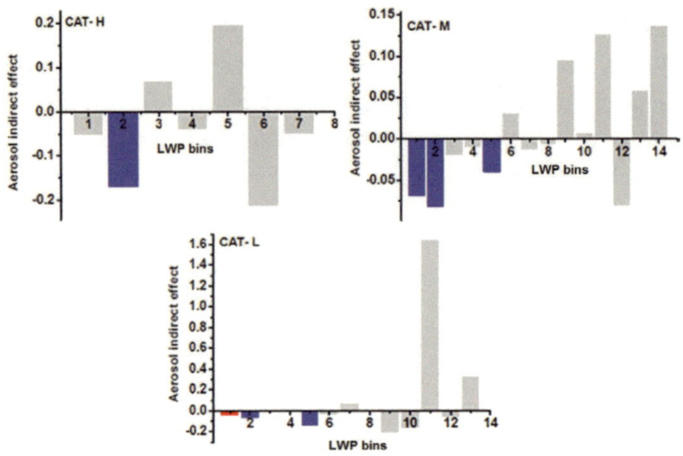

Figure 9. Aerosol indirect effect (AIE) in different LWP bins at different aerosol loading groups H, M, and L during 2000–2010. The range of bin number against LWP is as follows: (1) 1–25; (2) 26–50; (3) 51–75; (4) 76–100; (5) 101–125; (6) 126–150; (7) 151–175; (8) 176–200; (9) 201–225; (10) 226–250; (11) 251–275; (12) 276–300; (13) 301–325 and (14) 326–350 gm^{-2}. (Blue color bars represent AIE significant at 95% of confidence level, Red color bars represent IE significant at 90% of confidence level and rest are less significant).

Figure 9 illustrates the AIE for different LWP bins at each category during 2000–2010. It also reveals that during 2000–2010, the average values of AIE at each category (CAT-H, CAT-M, and CAT-L) were -0.04 ± 0.14, 0.01 ± 0.07, and 0.10 ± 0.48, respectively. About 71% of LWP bins predicted negative AIE at CAT-H, out of which 20% of the LWP bins were found to be statistically significant at 95% level. At CAT-M, about 57% of LWP bins exhibited negative AIE out of which about 38% of the LWP bins were significant at 95% level. AIE at CAT-L was observed to be less prominent as compared to other two categories. About 57% of LWP bins at CAT-L depicted negative AIE out of which 25 and 13% of LWP bins were significant at the level of 95 and 90%, respectively.

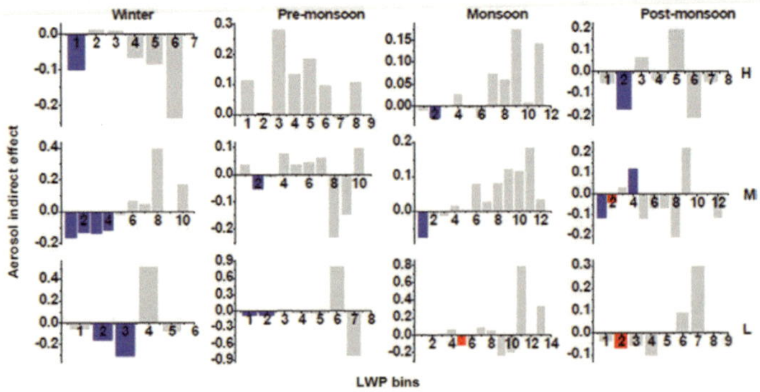

Figure 10. Same as in **Figure 9**, but for seasonal AIE. (Upper panel represents AIE for CAT-H, middle panel shows AIE for CAT-M and lower panel indicates CAT-L.) (Blue color bars represent AIE significant at 95% of confidence level, Red color bars represent IE significant at 90% of confidence level and rest are less significant).

Station	Winter	Pre-monsoon	Monsoon	Post-monsoon
DSL	−0.008 (60%)	0.141 (54.55%)	0.049 (53.85%)	−0.067 (63.64%)
MND	−0.143 (100%)	0.156 (85.71%)	0.118 (75%)	−0.068 (75%)
MSL	−0.045 (71.43%)	0.030 (77.77%)	0.043 (75%)	−0.146 (83.33%)
LDN	0.15 (80%)	0.13 (100%)	−0.023 (75%)	0.07 (80%)
PTL	−0.11 (100%)	0.09 (100%)	0.05 (55.55%)	−0.09 (50%)
MZR	0.10 (60%)	0.23 (83.33%)	0.04 (62.5%)	−0.04 (66.66%)
PUN	−0.121 (100%)	−0.091 (60%)	0.003 (55.56%)	0.024 (50%)
STR	−0.057 (72.72%)	−0.013 (50%)	−0.057 (63.63%)	−0.008 (75%)
KPR	0.038 (60%)	0.106 (83.83%)	−0.250 (60%)	−0.058 (85.71%)

Percentage of cases of positive/negative AIE are given in parenthesis.

Table 4. The seasonal average value of AIE for fixed LWP during 2000–2010.

Seasonal AIE at each category was estimated and is presented in **Figure 10** which shows that during winter, LWP bins at all the three categories demonstrated prominent anti-Twomey (negative AIE) effect, out of which 25, 80, and 50% of the LWP bins were significant at 95% level for CAT-H, CAT-M, and CAT-L, respectively. In pre-monsoon, LWP bins showed positive AIE at CAT-H (86%), and CAT-M (60%) while at CAT-L about 71% of LWP bins elucidated negative AIE. At CAT-H, CAT-M, and CAT-L, in monsoon, the percentage of LWP showing positive AIE were found to be about 60, 75, and 58%, respectively. In post-monsoon, about 71, 70, and 63% of LWP bins showed negative AIE for CAT-H, CAT-M, and CAT-L, respectively. Seasonal means of AIE for fixed LWP at each station were also estimated for the period of 2000–2010, and the results are given in **Table 4**. It reveals that majority of the stations showed the occurrence of negative AIEs during winter (about 67%) and post-monsoon (about 78%), respectively. However, on an average 78 and 67% of the stations revealed the occurrence of positive AIEs in pre-monsoon and monsoon seasons, respectively.

The estimation of AIE at each category and at each station indicates that the sign of AIEs changes with space and time (season). The main reason behind this positive or negative nature of the AIE could be linked with changes in circulation pattern and associated long-range transport of air masses containing different types of aerosols from the different sources [16]. For example, the mixing state of the black carbon, i.e., whether it is located in the interstitial aerosol, or included within the cloud droplets, can influence the resulting cloud droplet population and optical properties and the black carbon aerosol can change sign of the radiative forcing from negative to positive [6]. Additionally, dust plume existing under the cloud can influence the direct, indirect, and semi-direct effect [50]. However, there are several other factors along with meteorological conditions that could contribute to the observed changes in indirect effect [51].

5. Conclusions

MODIS satellite retrievals were used to investigate the spatiotemporal variations of MODIS retrieved aerosol and cloud products over Western Himalayan and Deccan Plateau regions. The data were also employed to study aerosol–cloud interaction and to quantify AIE over these regions for fixed LWP bins. The major conclusions of the present study are as follows:

- The analysis of MODIS retrieved aerosol and cloud products, the interrelation between cloud between aerosol parameters and the estimation of the AIE reveal that the aerosol–cloud relationship as well as AIE changes magnitude and sign both with space and time (season). The change in the atmospheric circulation patterns, in-flux/out-flux of different types of aerosols and varying prevalent meteorological conditions may be the reason behind this positive/negative aerosol indirect effect. Analysis of MODIS retrievals further revealed that the aerosols can change microphysical properties of clouds. However, it is evident that such changes are not determined by aerosols alone as there may be several contributing factors.
- At all the stations, during 2000–2010, the sign of AE-AOD correlation exhibited spatio-temporal variability. During winter, AE was positively correlated with AOD while during monsoon the correlation was negative with 95% of significance level at all the stations. In pre-monsoon, the majority of the stations (78%) showed inverse AE-AOD relationship, out of which about 86% of the correlations were significant at 95% confidence level. The changeover in the sign of AE-AOD correlation may be attributed to the hygroscopic growth of the aerosol particles producing a significant shift in aerosol size spectrum which in turn can substantially influence the magnitude of the AE.
- Monthly variation of AOD and LWP for the study period and their seasonal correlation analysis points out that the sign of correlation (negative/positive) changes from season to season and place to place indicating spatiotemporal variability in these correlations. Reduction in LWP with AOD during winter, pre-monsoon, monsoon, and post-monsoon seasons may be attributed to the warming of the clouds due to the dust aerosols. Dust aerosols absorb the incoming solar radiation resulting into increase in cloud layer temperature, thereby enhancing evaporation of cloud droplets leading to the reduction of LWP, i.e., the so-called semi-direct effect.
- Estimated values of AIE showed that the average values of AIE during 2000–2010, at each category (CAT-H, CAT-M, and CAT-L) were -0.04 ± 0.14, 0.01 ± 0.07, and 0.10 ± 0.48, respectively. About 57–71% of LWP bins revealed negative AIE at each category. During winter, prominent anti-Twomey effect (with 25–80% of significance level) was observed in all the three categories. In pre-monsoon, 86% (at CAT-H) and 60% (at CAT-M) of LWP bins showed positive AIE while about 71% (at CAT-L) of LWP bins exhibited negative AIE values.

http://dx.doi.org/10.5772/65237

Author details

Sandeep R. Varpe[1,3], Gajanan R. Aher[1*], Amol R. Kolhe[1,2] and Sanjay D. More[4]

*Address all correspondence to: aher.g.r@gmail.com

1 Physics Department, Nowrosjee Wadia College, Pune, India

2 Department of Physics, Savitribai Phule Pune University, Pune, India

3 International Institute of Information Technology, Hinjewadi, Pune, India

4 Department of Atmospheric and Space Sciences, University of Pune, Pune, India

References

[1] Tai APK, Mickley LJ, Jacob DJ. Correlations between fine particulate matter (PM2.5) and meteorological variables in the United States: implications for the sensitivity of PM2.5 to climate change. Atmospheric Environment. 2010; 44: 3976–3984.

[2] Lee SS, Penner JE. Dependence of aerosol-cloud interactions in stratocumulus clouds on the liquid-water path. Atmospheric Environment. 2011; 45: 6337–6346.

[3] Satheesh SK, Moorthy KK. Radiative effects of natural aerosols: a review. Atmospheric Environment. 2005; 39: 2089–2110.

[4] Kaufman YJ, Boucher O, Tanre D, Chin M, Remer LA, Takemura T. Aerosol anthropogenic component estimated from satellite data. Geophysical Research Letters. 2005; 32: L17804.

[5] Sundstrom AM, Arola A, Kolmonen P, Xue Y, de Leeuw G, Kulmala AM. On the use of satellite remote-sensing approach for determining aerosol direct radiative effect over land: a case study over China. Atmospheric Chemistry and Physics. 2015; 15: 505.

[6] Ramanathan V, Crutzen PJ, Kiehl JT, Rosenfeld D. Aerosols, climate and the hydrological cycle. Journal of Geophysical Research. 2001; 294: 2119–21124.

[7] Kaufman YJ, Koren I. Smoke and pollution aerosol effect on cloud cover. Science. 2006; 313: 655–658.

[8] Niu F, Li Z. Systematic variations of cloud top temperature and precipitation rate with aerosols over the global tropics. Atmospheric Chemistry and Physics. 2012; 12: 8491–8498.

[9] Houghton JT. et al editors. Climate Change. The scientific basis: contribution of working group I to the third assessment report of the Intergovernmental Panel on Climate Change (IPCC). Cambridge University Press, Cambridge (UK); 2001: 525–582.

[10] Myhre G, Stordal F, Johnsrud M, Kaufman YJ, Rosenfeld D, Storelvmao T, Kristjansson JE, Berntsen TK, Myhre A, Isaksen ISA. Aerosol-cloud interaction inferred from MODIS satellite data and aerosol global models. Atmospheric Chemistry and Physics. 2007; 7: 3081–3101.

[11] Penner JE, Dong XQ, Chen Y. Observational evidence of a change in radiative forcing due to the indirect aerosol effect. Nature. 2004; 427(6971): 231–234.

[12] Balakrishnaiah G, Raghavendra Kumar K, Suresh Kumar Reddy B, Rama Gopal K, Reddy RR, Reddy LSS, Nazeer Ahammed Y, Narasimhulu K, Krishnamoorthy K, Suresh Babu S. Analysis of optical properties of atmospheric aerosols inferred from spectral AODs and Angstrom wavelength exponent. Atmospheric Environment. 2011; 45: 1275–1285.

[13] Rosenfeld D, Lohmann U, Raga GB, O'Dowd CD, Kulmala M, Fuzzi S, Reissell A, Andreae MO. Flood or drought: how do aerosols affect precipitation. Science. 2008; 321: 1309–1313.

[14] Andreae MO, Rosenfeld D, Artaxo P, Costa A, Frank AGP, Longo KM, Silva-Dias MAF. Smoking rain clouds over the Amazon. Science. 2004; 303: 1337–1342.

[15] Ten Hoeve JE, Remer LA, Jacobson MZ. Impacts of water vapor/aerosol loading trends and land cover on aerosol microphysical and radiative effects on clouds during the Amazon biomass burning season. Atmospheric Chemistry and Physics (Discussions). 2010; 10: 24919–24960.

[16] Panicker AS, Pandithurai G, Dipu S. Aerosol indirect effect during successive contrasting monsoon seasons over Indian subcontinent using MODIS data. Atmospheric Environment. 2010; 44: 937–943.

[17] Lohmann U, Feichter J. Global indirect aerosol effects: a review. Atmospheric Chemistry and Physics. 2005; 5: 715–737.

[18] Storelvmo T, Kristjansson JE, Myhre G, Johnsrud M, Stordal F. Combined observational and modeling based study of the aerosol indirect effect. Atmospheric Chemistry and Physics. 2006; 6: 3583–3601.

[19] Ten Hoeve JE, Jacobson MZ, Remer LA. Comparing results from a physical model with satellite and *in situ* observations to determine whether biomass burning aerosols over the Amazon brighten or burn off clouds. Journal of Geophysical Research. 2012; 117: D0820319.

[20] Kumar A. Variability of aerosol optical depth and cloud parameters over North Eastern regions of India retrieved from MODIS satellite data. Journal of Atmospheric and Solar Terrestrial Physics. 2013; 100–101: 34–49.

http://dx.doi.org/10.5772/65237

[21] More SD, Pradeep Kumar P, Gupta P, Devara PCS, Aher GR. Comparison of aerosol products retrieved from AERONET, MICROTOPS and MODIS over a tropical urban city, Pune, India. Aerosol and Air Quality Research. 2013; 13: 107–121.

[22] King MD, Kaufman YJ, Menzel WP, Tanré D. Remote sensing of Cloud, Aerosol, and Water vapor properties from the Moderate Resolution Imaging Spectroradiometer (MODIS). IEEE Transactions on Geoscience and Remote Sensing. 1992; 30(1): 2–26.

[23] Levy RC, Remer LA, Dubovik O. Global aerosol optical properties and application to moderate resolution imaging spectroradiometer aerosol retrieval over land. Journal of Geographical Research. 2007; 112: D13210.

[24] King MD, Menzel WP, Kaufman YJ, Tanre D, Gao BC, Platnick S, Ackerman SA, Remer LA, Pincus R, Hubanks PA. Cloud and aerosol properties, precipitable water, and profiles of temperature and humidity from MODIS. IEEE, Transactions on Geoscience and Remote Sensing. 2003; 41: 442–458.

[25] Remer LA, Kaufman YJ, Tanre D, Matoo S, Chu DA, Martins JV, Li RR, Ichoku C, Levy RC, Kleidman RG, Eck TF, Vermote E, Holben BN. The MODIS aerosol algorithm, products, and validation. Journal of Atmospheric Sciences. 2005; 62: 947–973.

[26] Feingold G, Eberhard WL, Veron DE, Previdi M. First measurements of the Twomey indirect effect using ground-based remote sensors. Geophysical Research Letters. 2003; 30: 1287.

[27] Gautam R, Liu Z, Singh RP, Hsu NC. Two contrasting dust-dominant periods over India observed from MODIS and CALIPSO data. Geophysical Research Letters. 2009; 36: L06813.

[28] Prasad AK, Singh RP, Singh AA. Variability of aerosol optical depth over Indian subcontinent using MODIS data. Journal of the Indian Society of Remote Sensing. 2004; 32(4): 313–316. http://home.iitk.ac.in/~ramesh/jisrs/jisrs.html.

[29] Ramchandran S, Cherian B. Regional and seasonal variations in aerosol optical characteristics and their frequency distributions over India during 2001-2005. Journal of Geophysical Research. 2008; 113: D08207.

[30] Tripathi SN, Dey S, Chandel A, Shrivastava S, Singh RP, Holben BN. Comparison of MODIS and AERONET derived aerosol optical depth over the Ganga basin, India. Annals of Geophysics. 2005; 23: 1093–1101.

[31] Dey S, Tripathi SN, Singh RP, Holben BN. Influence of dust storms on the aerosol optical properties over the Indo-Gangetic basin. Journal of Geophysical Research. 2004; 109: D20211.

[32] Balakrishnaiah G, Kumar KR, Reddy BS, Gopal KR, Reddy RR, Reddy LSS, Swamulu C, Ahammed YN, Narasimhulu K, Krishnamoorthy K, Babu SS. Spatio-temporal variation in aerosol optical depth and cloud parameters over Southern India retrieved from MODIS satellite data. Atmospheric Environment. 2012; 47: 435–445.

[33] Mishra A, Klingmueller K, Fredj E, Lelieveld J, Rudich Y, Koren I. Radiative signature of absorbing aerosol over the eastern Mediterranean basin. Atmospheric Chemistry and Physics. 2014; 14: 7213–7231.

[34] Dubovik O, Sinyuk A, Lapyonok T, Sinyuk A, Mishchenko MI, Yang P, Eck TF, Volten H, Munoz O, Veihelmann B, van der Zander WJ, Sorokin M, Slutsker I. Application of light scattering by spheroids for accounting for particle non-sphericity in remote sensing of desert dust. Journal of Geophysical Research. 2006; 111: D11208. doi: 10.1029/2005JD006619.

[35] Aher GR, Pawar GV, Gupta P, Devara PCS. Effect of a major dust storm on optical, physical, and radiative properties of aerosols over coastal and urban environments in Western India. International Journal of Remote Sensing. 2014; 35: 871–903. doi: 10.1080/01431161.2013.873153.

[36] Graf HF. The complex interaction of aerosols and clouds. Science. 2004; 303: 1309–1311.

[37] Li Z, Niu F, Fan J, Liu Y, Rosenfield D, Ding Y. Long-term impacts of aerosols on the vertical development of clouds and precipitation. Nature Geoscience. 2011; 4(12): 888–894.

[38] Wright ME, Atkinson DB, Ziemba L, Griffin R, Hiranuma N, Sarah B, Lefer B, Flynn J, Perna R, Rappengluck B, Luke W, Kelley P. Extensive aerosol optical properties and aerosol mass related measurements during TRAMP/Tex AQS2006-Implications for PM compliance and planning. Atmospheric Environment. 2010; 44: 4035–4044.

[39] Philipp F, Markus F, Fritzsche L, Petzold A. Measurement of ultrafine aerosol size distributions by a combination of diffusion condensation particle counters. Journal of Aerosol Science. 2006; 37: 577–597.

[40] Walcek CJ. Cloud cover and its relationship to other meteorological factors during a springtime midlatitude cyclone. Monthly Weather Review. 1994; 120:1021–1035.

[41] Kaskaouties DG, Kharol SK, Sinha PR, Singh RP, Kambezidis HD, Sharma AR, Badarinath KVS. Extremely large anthropogenic-aerosol contribution to total aerosol load over the Bay of Bengal during the winter season. Atmospheric Chemistry and Physics. 2011; 11: 7079–7117.

[42] Chou MD, Chan PK, Wang M. Aerosol radiative forcing derived from Sea WiFS-retrieved aerosol optical properties. Journal of the Atmospheric Sciences. 2002; 59: 748–757.

[43] Indira G, Bhaskar BV, Krishnaswamy M. The Impact of aerosol optical depth impacts on rainfall in two different monsoon periods over Madurai, India. Aerosol and Air Quality Research. 2013; 13: 1608–1618.

[44] Yuan T, Li Z, Zhang R, Fan J. Increase of cloud droplet size with aerosol optical depth: an observation and modeling study. Journal of Geophysical Research. 2008; 113: D04201.

[45] Bhawar RL, Devara PCS. Study of successive contrasting monsoons (2001-2002) in terms of aerosol variability over a tropical station Pune, India. Atmospheric Chemistry and Physics. 2010; 10: 29–37.

[46] Huang J, Lin B, Minnis P, Wang T, Wang X, Hu Y, Yi Y, Ayers JK. Satellite-based assessment of possible dust aerosols semi-direct effect on cloud water path over East Asia. Geophysical Research Letters. 2006; 33: L19802.

[47] Peixoto JP, Oort AH. The Atmospheric Branch of the Hydrological Cycle and Climate. Street-Perrott A, Beran M, Ratcliff R, editors. Variations in the Global Water Budget. Springer Netherlands (D. Reidel Publishing Company) 1983. 5–65. DOI: 10.1007/978-94-009-6954-4_2.

[48] Ranjan RR, Joshi HP, Iyer KN. Spectral variation of total column aerosol optical depth over Rajkot: a tropical semi-arid Indian station. Aerosol and Air Quality Research. 2007; 7: 33–45.

[49] El-Askary H, Kafatos M. Dust storm and black cloud influence on aerosol optical properties over Cairo and the Greater Delta Region, Egypt. International Journal of Remote Sensing. 2008; 29(24): 7199–7211. doi:10.1080/01431160802144179.

[50] Su J, Huang J, Fu Q, Minnis P, Ge J, Bi J. Estimation of Asian dust aerosol effect on cloud radiation forcing using Fu-Liou radiative model and CERES measurements. Atmospheric Chemistry and Physics. 2008; 8: 2763–2771.

[51] Yu H, Fu R, Dickinson RE, Zhang Y, Chen M, Wang H. Inter-annual variability of smoke and warm cloud relationships in the Amazon as inferred from MODIS retrievals. Remote Sensing of Environment. 2007; 111(4): 435–449.